Buried Waste in the Seabed – Acoustic Imaging and Bio-toxicity

Results from the European SITAR Project

Philippe Blondel and Andrea Caiti

Buried Waste in the Seabed – Acoustic Imaging and Bio-toxicity

Results from the European SITAR Project

Springer

Published in association with
Praxis Publishing
Chichester, UK

PRAXIS

Dr Philippe Blondel, C.Geol., FGS, Ph.D., M.Sc.
Senior Scientist
Department of Physics
University of Bath
Bath
UK

Professor Andrea Caiti
Director
Interuniversity Centre of Integrated Systems for the Marine Environment
University of Pisa
Pisa
Italy

SPRINGER–PRAXIS BOOKS IN GEOPHYSICAL SCIENCES
SUBJECT *ADVISORY EDITOR*: Philippe Blondel, C.Geol., FGS, Ph.D., M.Sc., Senior Scientist, Department of Physics, University of Bath, UK

ISBN 10: 3-540-28120-7 Springer Berlin Heidelberg New York
ISBN 13: 978-3-540-28120-7 Springer Berlin Heidelberg New York

Springer is part of Springer-Science + Business Media (springer.com)

Bibliographic information published by Die Deutsche Bibliothek

Die Deutsche Bibliothek lists this publication in the Deutsche Nationalbibliografie; detailed bibliographic data are available from the Internet at http://dnb.ddb.de

Library of Congress Control Number: 2006932579

Printed in Germany

Cover design: Jim Wilkie
Project management: Originator Publishing Services, Gt Yarmouth, Norfolk, UK

Printed on acid-free paper

Contents

Foreword

Forty per cent of the world's population live within 100 km of the coastline and hundreds of millions of people are dependent on the oceans for their livelihood. This is true for Europe, where our seas and coastal zones are of strategic importance to all Europeans. They are home to a large percentage of our citizens, a major source of food and raw materials, a vital link for transport and trade, the location of some of our most valuable habitats, and the favoured destination for our leisure time. Yet Europe's seas are facing serious problems of habitat destruction, loss of biodiversity, contamination of water resources and resource depletion. In addition, the consequences of global climate change (e.g., sea level rise and increase in the frequency and intensity of storms) are likely to be increasingly felt in the future.

Despite these pressures, for many the seafloor may seem safe from the human disturbances that threaten other environments. Yet most natural and artificial wastes – such as sewage sludge, dredged spoils from harbours and estuaries, dangerous synthetic organic compounds and packaged goods – make their way to the seafloor over time.

The dumping of waste at sea, or "ocean dumping", is generally banned worldwide and is regulated by two conventions: the London Dumping Convention (1975) and the MARPOL[1] Convention (1978). The motivation for banning ocean dumping gained momentum when, for example, contaminated wastes from sewage-derived microorganisms were discovered at public beaches, and shellfish beds were contaminated with toxic metals. As a consequence, significant progress has been made in the prevention of ocean dumping.

However, less has been done to eliminate the effects of past dumping practice. A particular difficulty arises from the fact that the information on toxic dumpsites is not easily available, since, not surprisingly, many dumping operations involving

[1] MARPOL is short for MARine POLution.

hazardous material – of an industrial or military nature – have been carried out covertly. In areas where information is available, however, the problem is becoming of increasing concern.

One such area is the Baltic Sea, where existing documentation indicates that at least 65 000 tons of toxic chemical munitions have been dumped in these waters in the post-World War II years. The Helsinki Commission has investigated the status of Baltic Sea dumping, and it has recommended further *in situ* research on the toxicity and bioaccumulation effects on water, sediment and biota.

Starting from the above background, the SITAR project entitled "Seafloor Imaging and Toxicity: Assessment of Risk Caused by Buried Waste" (EC contract no. EVK3-CT-2001-00047) was established to make available instrumentation and systems to meet these requirements. The project started on 1 January 2002, had a duration of 3 years and brought together 10 partners from 6 European countries.

To meet its objectives the project required the merging of multidisciplinary scientific and technological expertise including: marine acoustics propagation and signal processing; biological and chemical toxicological analysis applied to marine biota; design, development and marketing of innovative marine instrumentation; image processing, geo-referenced data management and decision support systems; and environmental risk assessment and waste management. Such a diverse and complementary range of expertise could only be gathered together by combining competences that can be found in separate Europe-wide organizations. The SITAR project therefore provides a good example of how the "European added value" of the RTD[2] projects funded by the European Community can best be used to successfully address complex research problems.

The availability of the technology and scientific methodologies developed in SITAR will enable the implementation of risk prevention and monitoring measures prescribed by the European regional seas conventions, with potentially common procedures and approaches in all the European seas.

In the true spirit of the European Union's RTD Framework Programmes, the partners of the SITAR project through the publication of this book are disseminating their results to the widest possible community. The hope is that the potential users of these results (researchers, practitioners, local, regional and national authorities, to mention just a few) will test the validity of the methods and the tools that are presented and contribute constructively to their possible improvements for the benefit of the entire society.

In concluding, I would like to take this opportunity to thank all SITAR partners for their commitment, hard work and open collaborative spirit throughout the duration of the project.

Alan Edwards,
EC Scientific Officer for the SITAR Project
Brussels, December 2005

[2] RTD is short for Research and Technology Development.

Figures

Colour figures sit between pages 112/113

Tables

Abbreviations and acronyms

AChE	AcetylCholinEsterase
ACTC	ACetylThioCholine iodide
ARMINES	Association pour la recherche et le développement des processus inductriels
AUV	Autonomous Underwater Vehicle
B&K	Brüel & Kjaer
B[a]P	Benzo[a]Pyrene
BATH	University of Bath
BIE	Boundary Integral Equation
CHEMU	see HELCOM CHEMU
DBMS	Data Base Management System
DDT	DichloroDiphenylTrichloroethane
DE	Differential Evolution
DGPS	Differential Global Positioning System (Differential GPS)
DMF	DiMethylFormamide
DPC	Displaced Phase Centre
DSS	Decision Support System
DTNB	DiThiobisNitroBenzoate
dw	dry weight
ECAT	Environmental Centre for Administration and Technology
ECOD	Ethoxycoumarin-*O*-deethylase
EN	Ecole Navale
EROD	Ethoxyresorufin-*O*-deethylase
ESRI	Environmental Systems Research Institute
FARIM	Frequency Analysis-based Roughness and Impedance estimation Method
FM	Frequency Modulation
FOI	Swedish Defense Research Establishment

G	Gaussian probability density function (or Gaussian distribution)
G-G-R	Gaussian–Gaussian–Rayleigh
GA	Genetic Algorithm
GESMA	Groupe d'Études Sous-Marines de l'Atlantique
GIS	Geographical Information System
GPS	Global Positioning System
GUI	Graphical User Interface
HELCOM	Helsinki Commission
HELCOM CHEMU	ad hoc working group on dumped chemical munitions of the Helsinki Commission
HF	High Frequency
HPLC	High-Performance Liquid Chromatography
ICS	Inertial Coordinate System
IEC	International Electrotechnical Commission
ISACS	Integrated System for Analysis and Characterization of the Seafloor
ISME	Interuniversity Centre of Integrated Systems for the Marine Environment
ISO	International Organization for Standardization
IT	Inertial Tensor
KDA	Kongsberg Defense and Aerospace
KTH	Royal Institute of Technology, Stockholm
LBL	Long BaseLine
LF	Low Frequency
LSM	Least Squares Method
LU	Lower triangular times Upper triangular
MARPOL	MARitime POLution convention
MAS	Multiple-Aspect Scattering
MI	Mean Intensity
ML	Maximum Likelihood
MLE	Maximum Likelihood Estimator
MOM	Method Of Moments
MRF	Markov Random Field
MRU	Motion Reference Unit
NaCl	Sodium chloride
NADPH	Nicotinamide Adenine Dinucleotide PHosphate
NETCDF	NETwork Common Data Format
NTNU	Norwegian University of Science and Technology
OCS	Object Coordinate System
PAH	Polycyclic Aromatic Hydrocarbon (ΣPAH: sum of PAH)
PBB	PolyBrominated Biphenyl
PCB	PolyChlorinated Biphenyl (ΣPCB: sum of PCB)
PDF	Probability Density Function
Plums	*Pl*atform for *u*nderwater *m*easurement *s*ystems

P-SAS	Planar Synthetic Aperture Sonar
PSSS	Parametric Side-Scan Sonar
R	Rayleigh probability density function (or Rayleigh distribution)
RK	Ray–Kirchhoff
RMS	Root Mean Square
ROV	Remotely Operated Vehicle
RT90	Rikets Nät 1990 (Swedish: Swedish grid). Coordinated system used for government maps in Sweden
Rx	Receiver
SAS	Synthetic Aperture Sonar
SD	Standard Deviation
SEPA	Swedish Environmental Protection Agency
SGU	Swedish Geological Survey
SU	Stockholm University
TL	Transmission Loss
TNT	TriNitroToluene
TOC	Total Organic Carbon
TOPAS	Parametric sub-bottom profiler (Kongsberg)
TVG	Time-Varying Gain
Tx	Transmitter
UV	UltraViolet
VIS	VISual
VRML	Virtual Reality Modeling Language
W	Weibull probability density function (or Weibull distribution)
WGS84	World Geoïdal Surface 1984
WP	Work Package
ww	wet weight

Contributors

(in alphabetical order)

Gun ÅKERMAN
SU, Sweden

Lennart BALK
SU, Sweden

Philippe BLONDEL
BATH, UK

Andrea CAITI
ISME, Italy

Mario COSCI
BATH, UK and University of Pisa, Italy

Peter F. DOBBINS
BATH, UK

Johnny DYBEDAL
KDA, Norway

Alan EDWARDS
EC Scientific Officer, Brussels, European Union

Lina GELAZIENE
ECAT, Lithuania

Kerstin GRUNDER
SU, Sweden

Tomas HANSSON
SU, Sweden

Jens M. HOVEM
NTNU, Norway

Nisabha (Nic) JAYASUNDERE
BATH, UK

Per JONSSON
SEPA, Sweden

Ilkka KARASALO
FOI, Sweden

Magne A. LARSEN
NTNU, Norway

Per MOREN
FOI, Sweden

Vittorio MURINO
ISME, Italy

Nicholas G. PACE
BATH, UK

Maria PALMESE
ISME, Italy

Michela ROBBA
ISME, Italy

Roberto SACILE
ISME, Italy

Patrik SKOGQVIST
FOI, Sweden

Henrik SUNDBERG
SU, Sweden

Ulla TJÄRNLUND
SU, Sweden

Andrea TRUCCO
ISME, Italy

Manell ZAKHARIA
ARMINES/EN, France

Yngve ZEBÜHR
SU, Sweden

SITAR project partners

This book reports activities and results that have taken place within the SITAR project (funded by the European Union). Scientific activities were divided among the partners according to the specific expertise of each. In the book the scientific teams from each project partner are collectively indicated through the partner acronym. The acronyms used, together with the scientific coordinator and main point of contact (p.o.c.) for each partner, are listed in alphabetical order:

ARMINES/EN

Association pour la recherche et le développement des méthodes et processus industriels, and Naval Academy Research Centre; École Navale, Lanvéoc Poulmic, 29240 Brest Naval, France
p.o.c.: Manell Zakharia, *zakharia@ecole-navale.fr*

BATH

Department of Physics, University of Bath, Claverton Down, BA2 7AY Bath, United Kingdom
p.o.c.: Philippe Blondel, *pyspb@bath.ac.uk*

ECAT

Environmental Centre for Administration and Technology – Lithuania, Lydos str. 4, 3000 Kaunas, Lithuania
p.o.c.: Lina Gelaziene, *lina@ecat.lt*

FOI

Swedish Defense Research Establishment, Enköpingsvägen 126, 172 90, Stockholm, Sweden
p.o.c.: Ilkka Karasalo, *ilkka@foi.se*

ISME

Interuniversity Centre of Integrated Systems for the Marine Environment, c/o DIST, University of Genova, Via Opera Pia 13, 16145 Genova, Italy
p.o.c.: Andrea Caiti, *caiti@dsea.unipi.it*

KDA
Kongsberg Defense and Aerospace, Sensors and Systems, Strandveien 1, 7501, Stjørdal, Norway
p.o.c.: Johnny Dybedal, *Johnny.dybedal@kongsberg.com*

KTH
Royal Institute of Technology, Stockholm, Marcus Wallenberg Lab., Teknikringen 8, 10044 Stockholm, Sweden
p.o.c.: Anders Nillsson, *andersni@fkt.kth.se*

NTNU
Norwegian University of Science and Technology, Dept. Telecommunications O.S. Bragstads plass 2B, 7491 Trondheim, Norway
p.o.c.: Jens M. Hovem, *hovem@iet.ntnu.no*

SEPA
Swedish Environmental Protection Agency, 10691 Stockholm, Sweden
p.o.c.: Per Jonsson, *Per.jonsson@environ.se*

SU
Inst. of Applied Environmental Research, Stockholm University, Frescativaegen 54a, 10691 Stockholm, Sweden
p.o.c.: Lennart Balk, *Lennart.balk@itm.su.se*

Part I

The SITAR Project: background, goals, project structure

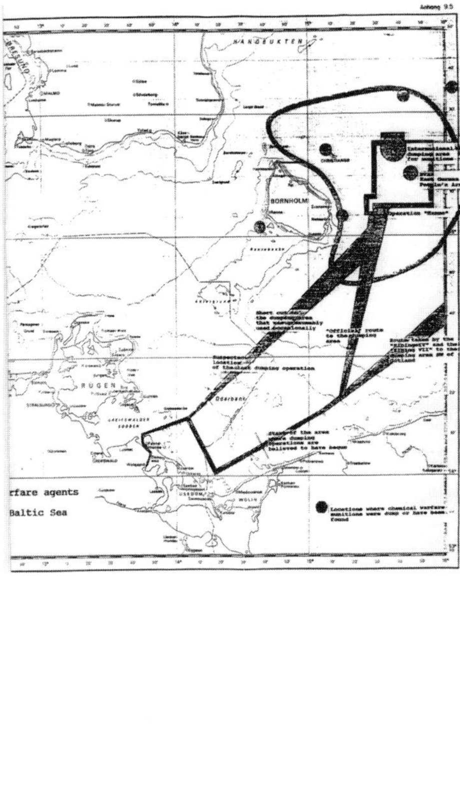

1

The SITAR Project: background, goals, project structure

A. Caiti

1.1 BACKGROUND: DUMPING AT SEA, EUROPEAN LEGISLATIONS AND ASSESSMENT OF BURIED WASTE

Dumping at sea is defined and regulated by two global conventions: the London Dumping Convention (1975), dealing with direct disposal of dumping at sea, and the MARPOL Convention (1973, modified 1978), dealing with ship-borne operations. At European level, the directives of the global conventions are included in several regional conventions which deal with marine pollution at a regional or individual sea basis (e.g., the Barcelona Convention for the Mediterranean Sea, 1975; the Helsinki Convention for the Baltic Sea, revised 1992; the Paris Convention for the North East Atlantic, 1992). In all these conventions, the signing countries have agreed "to prevent and eliminate" pollution at sea from various sources, including dumping; moreover, the conventions apply the precautionary principle – that is, "the enforcement of preventive measures when there is reason to assume that substances or energy introduced into the marine environment may create hazards to human health, harm living resources and marine ecosystems even when there is no conclusive evidence of a causal relationship between inputs and their alleged effects."

As far as dumping at sea is concerned, the above conventions have allowed significant results to be obtained in terms of prevention and regulation of current dumping practice; however, much less has been done to eliminate or at least to monitor the effects of past dumping practice, in particular of toxic material. There have been several reasons for this lack of action. Among them, the fact that the information on toxic dumpsites is not easily available: not surprisingly, many dumping operations involving hazardous material, of industrial or military nature, have been done covertly, even before the signing of the London and MARPOL Conventions. In areas where information is available, however, the problem is becoming of increasing concern. One such area is the Baltic Sea, where existing documentation indicates that at least 65,000 tons of toxic chemical munitions have been dumped in the post-World War II years, with toxic agents including mustard and viscous mustard gas, adamsite, clark, lewisite and tabun. Another European area where similar dumpings have occurred is the North Sea (Rapsch, 2000); the Adriatic Sea is also suspected to be a recipient of toxic waste, although in this case there is a consistent lack of precise information. The Helsinki Commission – HELCOM – has conducted a specific study on the status of Baltic Sea dumping (HELCOM CHEMU, 1994), showing that the preservation status of the chemical containers range from well preserved to totally corroded; moreover, among the contaminants, mustard gas remains in stable form on the seabed, while the other contaminants react in water, producing arsenic compounds that also rest at the seabed. While ruling out the possibility of environmental damage on a large scale, the HELCOM working group has recommended further investigations, in particular on the following topics:

(1) further search for location of chemical waste, and in particular determination of the amount of toxic waste buried within the sediment, and determination of the state of corrosion of the containers;
(2) *in situ* investigation of the toxicity and bioaccumulation effects on water, sediment and biota due to leakage of toxic material.

Implementation of the above recommendations have been so far frustrated: for the first topic, because of the lack of adequate instrumentation for imaging of buried objects of the expected dimensions; for the second topic, due to the lack of appropriate analytical methods able to evaluate the *in situ* toxic impact on marine biota of a contaminated sediment, including bioaccumulation effects.

In particular, standard site surveys over a dumpsite (or a shipwreck site) are nowadays performed by first exploring the site with a side-scan sonar instrument, to collect a map of location of objects over the area of interest, and subsequently by inspection of each object with video-cameras operated from an ROV (Remotely Operated Vehicle). In this way it is possible to gather information on the nature of the objects, dimensions, orientation, state of preservation, etc. All this information is necessary in order to precisely evaluate means, costs and efforts required for future actions at the site (like, for instance, clean-up or recovery operations). While current instrumentation is effective for inspection of objects exposed on the seabed, neither state-of-the-art side-scan sonar systems nor, clearly, video-cameras are able to gather the same information as for objects buried within the seabed sediment. Standard sub-bottom profilers are often used to collect information on the subsurface structure, but these instruments operate at relatively low frequency, in order to ensure bottom penetration, and hence they do not possess the resolution needed to image buried objects of small dimensions (throughout the book by "small dimension" we mean a cylinder-like object of 1-m length and 0.1-m diameter).

As for environmental risk at a toxic dumpsite, a crucial role is played by the nature of the toxic contaminant. Previous studies have been conducted only for a few components found in known dumping areas, such as those of the Baltic Sea. These studies (Federal/Länder, 1993; HELCOM CHEMU, 1994 and references therein) have focused entirely on acute toxicity. This is of concern, since some compounds (such as organic hydrophobic structures) do not have direct acute toxic effects. Instead, they express their effects at a time long after exposure or during a prolonged exposure period. Stable organic pollutants also show the ability to bioaccumulate both by bioconcentration directly from the water phase, and by biomagnification via the food web. In many cases, the parent compounds are not toxic, while the ultimate toxic substances are formed during the metabolism of these compounds. Hence, acute toxicity investigations may provide only a limited and partial picture of the environmental status of a given area.

1.2 THE SITAR PROJECT: GOALS AND PARTNERSHIP

Starting from the above background, a team of ten laboratories and institutions Europe-wide (see p. xxiii) submitted in 2001 a project proposal to the European Union, in accordance with the rules and priorities established by the 5th Framework Programme for Research and Development, on the investigation of novel technologies and scientific methodologies for risk assessment of dumpsites in presence of buried waste containers. One of the most crucial aspects in any proposal preparation is the choice of project title, in order to produce an effective (and easy to spell) project acronym; after several tests, Philippe Blondel from the University of Bath proposed "Seafloor Imaging and Toxicity: Assessment of Risk caused by buried waste – SITAR." Once the project got approval from the European Union, Peter Dobbins, also from the University of Bath, designed the project logo.

SITAR was established with the goal of investigating scientific methodologies and techniques to make instrumentation and systems available, as first priority, to meet the requirements of the HELCOM recommendations, identified as the more detailed specifications of actions to assess the threat posed by toxic dumpsites. The project started on January 1, 2002, and had a duration of 3 years. The specific goals of the project were detailed as follows:

- developing acoustic methods and instrumentation for imaging of waste barrels/containers of small dimension buried in unconsolidated sea sediments;
- developing biological testing methods to determine the relative *in-situ* bioaccumulated toxicity at a contaminated site;
- integrating and making accessible the acoustical imaging data and the biotoxicological information to end-users and decision makers.

The research consortium (see p. xxiii) assembled to meet the goals included the necessary merging of multidisciplinary scientific and technological expertise: underwater acoustics, signal processing, marine biotoxicity, geo-referenced data management, decision support systems, environmental risk assessment and waste management. Such a diverse and complementary range of expertise could be gathered together only by combining competences found in separate organizations Europe-wide; cooperation at European level has allowed the critical mass to be established, in terms of human capital and resources, needed to address the challenges posed by the project's goals.

1.3 SPECIFIC OBJECTIVES AND METHODOLOGIES

In order to reach the goals, on the basis of the scientific state-of-the-art in 2001, the following detailed objectives were identified:

- investigation and test of the *parametric side-scan sonar*, a conceptually innovative sonar system based on the principle of bottom-penetrating parametric sonar, but with the imaging capabilities of traditional side-scan sonar. Data from the instrument were to be processed by generalizing the synthetic aperture sonar principle to 3D geometry, in order to obtain 3D high-definition acoustical images of the sub-bottom from a survey platform, or a tow fish, in navigation;
- investigation and experimental verification of *multiple-aspect scattering* measurement techniques, and related automatic data-processing strategies. The method relies on real-time adaptive geometrical configuration of bistatic transmitting/receiving configuration;
- investigation and test in the field of the technique of *nanoinjection* into fertilised fish eggs of extracts of seafloor sediment collected *in situ* at a dumpsite. Observation of the percentage of abnormally reproduced fish is intended to give a relative indication of bioaccumulated toxicity;
- development of a *GIS-integrated data presentation system* (geographical information system), including a decision support system (DSS) to improve and automatically analyse the acoustic images of the marine sub-bottom.

To provide a better understanding of some of the background ideas behind the identified objectives, an ideal description of a survey and inspection mission over a toxic dumpsite with SITAR methodologies is now given. It is assumed that the dumpsite location is approximately known, from previous surveys or from historical information. The first mission step would consist in site survey with the parametric side-scan sonar instrument, to identify buried (and exposed) objects of unknown type over the survey area. Once object position is determined from the 3D parametric side-scan survey, the multiple aspect scattering measurement method will serve the purpose of inspection of the buried objects, to provide data for their characterization and classification. In this sense, the two systems aim at reproducing the usual procedure for location and inspection of exposed objects at the seabed, where a standard side-scan sonar is used to determine where the potential objects are, and video-camera inspection is used to investigate the objects once their locations are known. In SITAR, the bottom-penetrating parametric side-scan is the counterpart of the traditional side-scan sonar, while multiple aspect-scattering measurements are used to replace video-cameras in order to get full 3D images of the buried objects from various orientations. In addition to acoustic inspection, bottom sediment samples are collected for bio-toxicological analysis over the survey area. After processing of both the acoustic data and the toxicological data, the results are merged in a geo-referenced database. The database gives the user access to the collected measurements at various levels of detail, from aggregated information as maps of buried/exposed objects, to inspection of individual containers or bottom samples. From these data, the user can make decisions on the general environmental risk of the area, and can assess the feasibility of clean-up operations involving recovery of the buried containers.

The decision-making process in itself was beyond the scope of SITAR. SITAR was supposed to provide the end-user with the tools and information from the dumpsite needed to coming to a rational and motivated decision. It has to be remarked that additional information is also needed by the user, as inspection of exposed waste containers (a dumpsite in general comprises both exposed and buried containers), and oceanographic information on the site. This additional information is obtainable with state-of-the-art instrumentation, and so was not investigated in itself. Nevertheless, the GIS system has been designed taking into account the needs of integrating this additional information as well. It has also to be remembered that SITAR was a research project, aimed at a "proof-of-concept" demonstration, and not at the realization of commercial equipment or at the conduction of a full risk assessment case study.

1.4 PROJECT STRUCTURE

The organizational structure of the project has been in work packages (WPs), each one subdivided into tasks. There have been five WPs in SITAR: the first three were devoted to buried object acoustical imaging (WP-1), bioaccumulated toxicity testing (WP-2) and data integration (WP-3). WP-4 and WP-5 were dedicated, respectively, to field experiment preparation and execution, and to project management. A conceptual block diagram of the data flow through the various WP activities is reported in Figure 1.1.

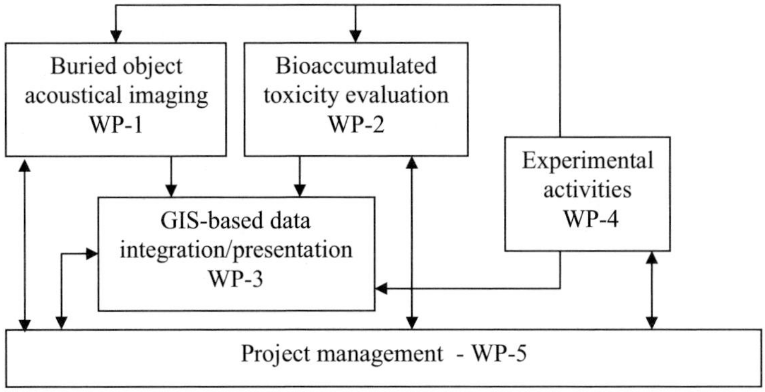

Figure 1.1. Conceptual data flow and relations among the project WPs.

From the point of view of the responsibility structure within the project, each WP had a WP leader: BATH (WP-1), SU (WP-2), ISME (WP-3). FOI has been the responsible partner for the experimental activities (WP-4), and ISME for project management (WP-5), including data and exploitation management, dissemination and overall project coordination. Other partners on WP-1 were ARMINES/EN, NTNU, KDA, FOI, KTH; the other partner on WP-2 was SEPA; other partners in WP-3 were ECAT and SEPA. All partners participated in the experimental and management activities.

1.5 STRUCTURE OF THE BOOK

This book summarizes the scientific activities and the achievements of SITAR in the three years of its activities. It is divided in four main parts, corresponding, respectively, to Methods employed, project Experimental activities, Results obtained applying the developed methods to the experimental data, Evaluation of the results with respect to the overall project goal – that is, an improvement in risk assessment of seabed dumpsites.

In particular, the Methods part focuses on the parametric side-scan sonar (PSSS) design and synthetic aperture signal processing (Section 2.1), the multiple-aspect scattering (MAS) measurement requirements and data processing (Section 2.2), the development of fast and accurate computational models for data simulation and interpretation (Section 2.3), the procedures for biotoxicological analysis (Section 2.4), the 3D acoustic image-processing techniques developed to extract from MAS data the relevant object features and to provide virtual rendering of the object (Section 2.5), the design and implementation of the GIS-based system (Section 2.6).

Experimental activities have been conducted partly as tank tests in controlled conditions (Section 3.1 – MAS; Section 3.2 – PSSS), partly in the field over a known and historically well-documented dumpsite in the Stockholm Archipelago, Baltic Sea (Section 3.3 – biotoxicological measurements; Section 3.4 – acoustic measurements). Experimental testing of SITAR techniques in the field did not have the purpose of conducting a case study of the given dumpsite, but of verifying in realistic conditions the feasibility of the SITAR techniques and procedures. As is too well known among marine scientists and engineers, and all experimental scientists working in harsh and extreme environments, experimentation at sea also has the drawback of preventing most of the time a systematic evaluation of the techniques under study, because of adverse sea conditions, limited experimental time, equipment faults, lack of or difficulties in obtaining ground truth data for comparison with experimental findings. Within SITAR, the availability of laboratory tank facilities has provided an additional means for systematic evaluation of the acoustic methodologies. Moreover, the experimental site has been purposely chosen as one for which a wealth of background data were available and obtainable by the project partners. This has not prevented the occurrence of malfunctioning, unexpected situations, ambiguous data interpretation, but has at least mitigated their influence in assessing the results and evaluating the project findings.

Data processing and results from the experimental activities are separately reported as: geoacoustic inversion at the experimental site, aimed at obtaining from the acoustic data the detailed acoustic description of the surficial part of the seabed (Section 4.1); PSSS data (Section 4.2); MAS technique (Section 4.3); biotoxicity measurements with the nanoinjection technique (Section 4.4); 3D processing of MAS acoustical images (Section 4.5). It has to be remarked that biotoxicity measurements have given – as results – that no particular toxicity can be detected in the area. This raises the potential question whether the area is effectively unpolluted (because pollutants are still safely confined in containers, for instance, or because they have leaked and dispersed long ago), or whether the investigated method is insensitive to toxicity. Additional field data from the Stockholm Archipelago have been made available to the SITAR team, and they are reported as well in Section 4.4, in which the same technique is applied and for which the method is indeed showing sensitivity to polluting factors.

Evaluation of the results from the point of view of application to seabed dumpsite risk assessment is conducted for the GIS-based data integration/presentation system (Section 5.1), the nanoinjection technique (Section 5.2), and acoustic methods (Section 5.3). It has to be remarked that the evaluation of the PSSS instrument in Section 5.3 relies heavily on GIS system capabilities of presenting in an integrated way the historical background data (the "ground truth") and the SITAR-generated data.

A Conclusion section, summarizing the work and connecting SITAR scientific achievements with the environmental needs that motivated the project, terminates the book.

1.6 READER'S GUIDE IN BRIEF

The book is addressed at educated scientists and engineers working in the research and development field. As such, it requires familiarity of the reader with basic aspects of underwater acoustics, marine biology, signal processing and computer science. The interdisciplinarity of the problem motivating the research, however, requires an effort from a reader trained in one specific discipline to approach and to also critically evaluate the results obtained in the other fields that make up the overall SITAR picture. For this reason, the intention of the various contributions is to make research results accessible, in their general form, to readers who are not researchers or specialists of the field as well. However, the book also includes the level of detail required by specialists; it has been an additional goal of the contributors to guarantee wide accessibility while maintaining scientific rigour and precision. Whether this goal has been reached, totally or partially, is for the reader to judge. The chapters of the book have been written whenever possible as self-contained scientific reports; however, it is clear that the motivations and the context of the various activities are within the broader scope of the SITAR project. References to each chapter are reported at the end of the book; while this may create some repetition in referencing, it is hoped that this will give the reader a better overall view of the research described in the book.

Part II

Methods

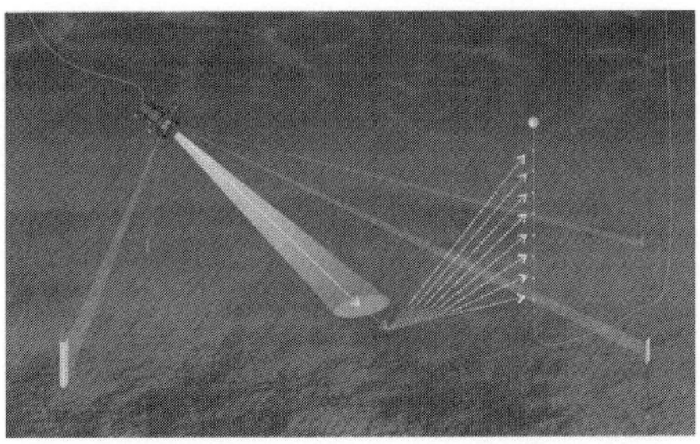

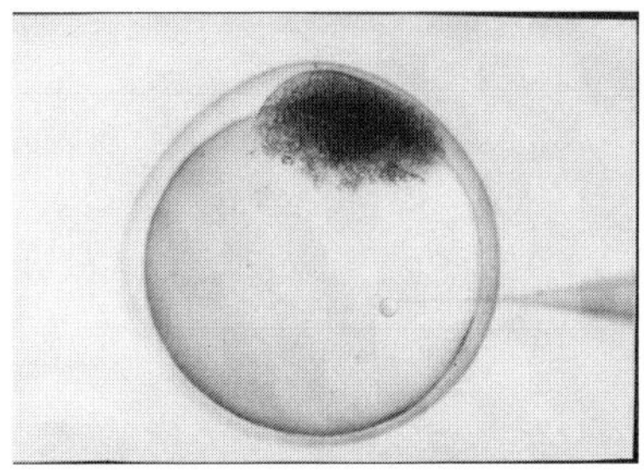

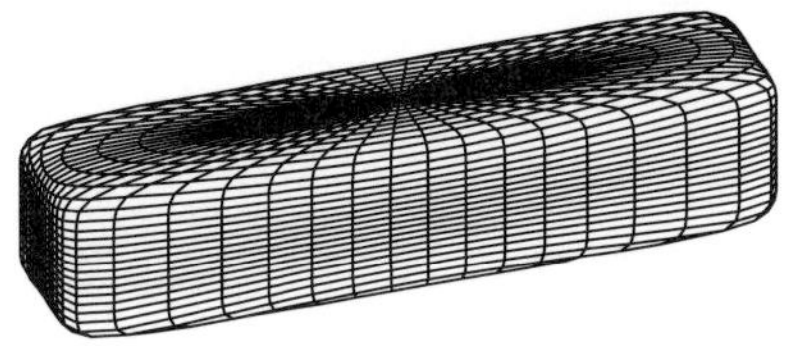

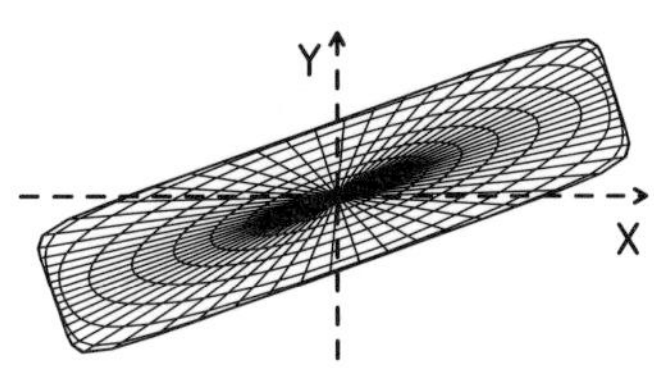

2

The parametric side-scan sonar instrument and synthetic aperture sonar processing

M. Zakharia and J. Dybedal

2.1 INTRODUCTION

Parametric and synthetic aperture sonar (SAS) techniques are commonly considered as competing techniques as they both aim to obtain "virtual" low-frequency arrays with a small angular aperture. The parametric technique is commonly associated with an echosounder-like geometry, and the SAS with a side-scan sonar configuration. They are both used for high-resolution sub-bottom imaging needed for imaging "objects" – for instance, in a dumpsite. The use of low frequencies is imposed by bottom penetration. Both techniques make use of small-size physical arrays (wide angular aperture) and compensate for the insufficient aperture either by nonlinear effects or by processing. The objects investigated in the SITAR project range in size from a dense aggregation of small objects of a few tens of centimetres up to isolated containers of about a few metres. As dumping may have occurred more than 50 years ago, they have been covered with unconsolidated sediments with a burial depth of less than a metre.

In the SITAR project, the simultaneous use of both parametric and synthetic techniques was investigated. The proposed sonar architecture employs parametric generation to obtain a narrow, bottom-penetrating beam. It compliments it with synthetic processing, to improve azimuthal resolution, and with dynamic focusing, to keep it constant with range.

2.2 PARAMETRIC SIDE-SCAN SONAR (PSSS)

Conventional sonar and echosounder systems are based on the linear behaviour of the transmitted acoustic signal. This means that the acoustical output signal has the same frequency as the electrical input signal, that the level changes linearly with the electrical input level and that the beamwidth is primarily inversely proportional to the dimensions (to first order) of the transducer array for these systems. In order to increase the spatial resolution for a given sonar system, the beamwidth should be minimized and the bandwidth should be maximized. The first requirement implies that the size of the transducer array may be very large and, consequently, the system may be inconvenient to handle and operate, especially when the output frequencies are low (<50 kHz). Wide bandwidth is now more easily available in linear systems due to new transducer materials and designs.

An alternative method for generating narrow-beam, wide-band acoustic signals is to utilize a parametric array which is based on the nonlinear propagation of high-intensity pulses in the water column. Nonlinear acoustics has been known for a long time and was first described by Rayleigh. Westervelt (1963) presented the idea of generating low-frequency sound with higher directivity than could be achieved by conventional sources by the use of a nonlinear propagation effect in water. An acoustic source based on this principle is often called a parametric array or a parametric antenna. The nonlinear behaviour occurs due to the fact that the velocity in any part of the acoustic wave is a function of the local pressure. The high-pressure part of the wave propagates faster than the low-pressure part. Eventually, after a certain distance a shockwave forms if the attenuation is low. For a sinusoidal wave, the different propagation velocity results in deformation of the wave and generation of harmonics. When a primary wave is composed of two high-frequency waves, f_1 and f_2, transmitted at high power, some of the energy is converted into waves at frequencies $2f_1, 2f_2, f_1 \pm f_2$, etc.

The most interesting frequency to be used in sub-bottom profiling is the difference frequency, $|f_1 - f_2|$, which is also called the secondary frequency. Typically, less than 1% of the energy is converted into the secondary frequency waveform. However, this is sufficient to achieve very good performance for the parametric sonar. Achievable relative bandwidth for the secondary wave may be as high as 100%, and the beamwidth may be relatively close to the beamwidth of the primary frequency signal.

In the SITAR project, a low-frequency side-scan sonar was designed for resolving and detecting small, buried objects with dimensions down to 0.2 m. The frequency range of interest is from 2 to 20 kHz and the beamwidth should be in the order of degrees. Unfortunately, such requirements would result in a very large conventional transducer array. A parametric array was therefore selected. The array has been designed to be also used as a conventional side-scan sonar at the 100-kHz primary frequency. Physical dimensions of the array are approx. 1.5×0.1 m (Figure 2.1). The measured approximate beamwidths alongtrack are 0.7° for the primary 100-kHz signal and 1.4° for the secondary 20-kHz signal. Typical transmitting source level measures for the sonar are 218 and 175 dB re 1 μPa @ 1 m for the primary and secondary frequencies, respectively. Two separate receiver line arrays were used, one placed above and the other below the transmitting transducer. Each receiver array comprised eight individual channels, and both the primary and the secondary frequency signals were logged on disk from each channel. Data from the two receiver arrays are used for interferometric depth estimations of the return echoes.

Figure 2.1. PSSS transducer array.

2.3 SYNTHETIC APERTURE SONAR (SAS)

Synthetic aperture processing for seabed imaging has seen a renewed interest during recent years in both civilian and military applications (see the special issue of the IEEE *Journal of Oceanic Engineering*, January 1992). Towards the end of the 1980s, several prototype systems were developed. SAS was mainly applied to side-scan sonar, but

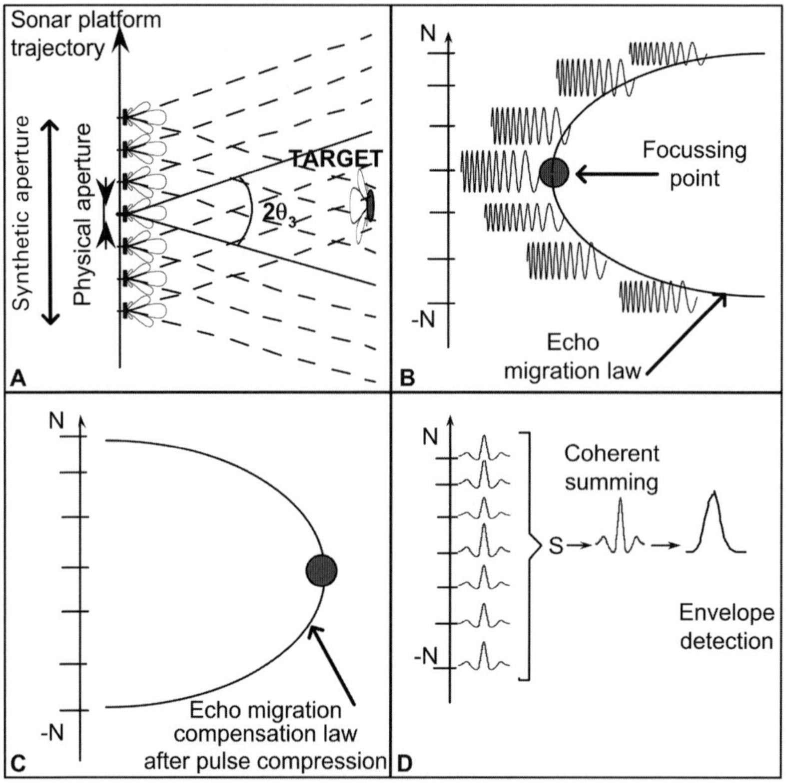

Figure 2.2. Principle of SAS.

could also be applied to multibeam or to planar arrays (Belkacem *et al.*, 2006). When the azimuthal resolution of a side-scan sonar is not sufficient, the echo of a given sonar target will be present in several consecutive pings (due to the fact that the main lobe of the transducer used – transmitter and/or receiver – is not narrow enough, see Figure 2.2A). SAS processing takes advantage of echo redundancy in order to build up a virtual array with a better resolution. Synthetic aperture consists, in fact, in building up a filter matched to the sequence of echoes issued from every target (Figure 2.2B):

$$SEQU = \sum_{-N}^{N} E_n(t)$$

Each echo of the sequence associated with a target can be written as:

$$E_n(t) = a_n \cdot F_A[F_T(s(\eta_n(t - \tau_n)))]$$

where:

- $s(t)$ is the transmitted signal;
- F_A is a filter associated with the directivity pattern of both transmitting and receiving arrays;

- F_T is a spatial filter associated with the impulse response (monostatic or bistatic depending on the sonar geometry) of the target for a given aspect angle (α_n);
- a_n is a normalizing constant including the transmission loss;
- η_n is the Doppler effect associated with v_n, the projection of the sonar velocity v on the sonar–target axis: $\eta_n = \dfrac{1 - v_n/c}{1 + v_n/c}$; and
- τ_n is the delay corresponding to the slant range R_n: $\tau_n = \dfrac{R_n}{c}$.

For a constant velocity along the x-axis (and a periodic transmitting rate, PRF), the expression of echo delay τ_n (hyperbola) can be given by simple geometrical considerations:

$$\tau_n^2 = \frac{4}{c^2}\left[A\left(\frac{n}{PRF}\right)^2 + R_0^2\right]$$

where $A = v^2$ and R_0 is the minimum slant range of the target.

SAS processing cannot be achieved in a general case, as the target attitude as well as its impulse responses are not known *a priori*. Several assumptions have to be made in order to make processing possible:

- F_A: the angular variation (that determines N, the sum bounds) is limited to the 3-dB aperture of the arrays. For simplification, the pattern is assumed constant in this range.
- F_T: the target is assumed to be small enough (compared with the wavelength) so it can be considered omnidirectional. F_T becomes an all-pass filter for all angles.
- a_n: the maximum echo migration ($R_N - R_0$) is assumed to be negligible as compared with the minimum range R_0. The transmission loss (TL) can thus be compensated by conventional TVG (time-varying gain).
- η_n: under the assumption that the platform trajectory is rectilinear with some small perturbations that could be measured or estimated (autofocusing techniques), the towing speed is slow enough that it can be ignored (or, at least, constitutes a second-order error). Optimal signal design can also be used to reinforce this hypothesis (using wideband Doppler-tolerant signals): no bias on the ambiguity function of the transmitted waveform, small contrast loss due to the Doppler effect (Châtillon *et al.*, 1992).

Under these assumptions, the sequence of echoes can be simplified:

$$SEQU = \sum_{-N}^{N} E_n(t - \tau_n)$$

In such a case (Zakharia and Châtillon, 2002), SAS processing will consist in compensating echo delay (or migration) (Figure 2.2C), adding all the echoes (after pulse compression) and computing the output envelope (Figure 2.2D). This processing has to be preceded by a filter matched to a single transmitted signal (standard matched filter for pulse compression). This simplified processing is

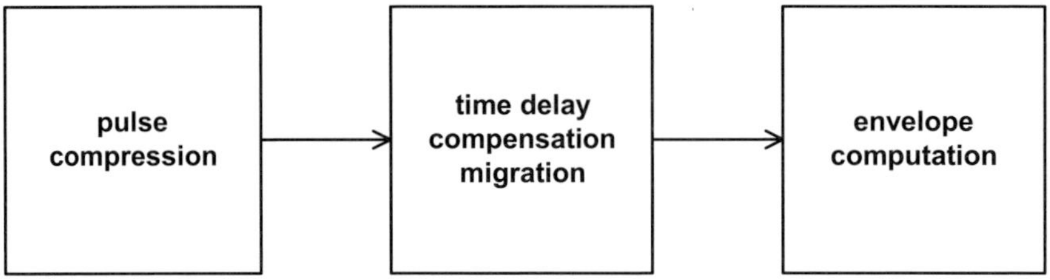

Figure 2.3. Principle of simplified wideband processing (Châtillon *et al.*, 1992).

described in Figure 2.3. Processing complexity is reduced considerably: instead of one filter matched to every position, the same filter is used for all positions. Apart from the pulse compression filter, SAS processing is now reduced to a simple compensation of echo migration delays.

In the conventional case, the angular aperture (at -3 dB) of the synthetic array obtained after focusing is equal to:

$$\theta_{sa} = \frac{\lambda_0}{2L_{sa}}$$

where λ_0 is the wavelength corresponding to the central frequency and L_{sa} is the length of the synthetic array. Processing will be extended to the parametric generated waves.

In order to avoid the range-sampling problem, dynamic focusing can be used and the length of the synthetic array can be adapted to the target range. Processing is commonly restricted to the -3-dB main lobe of the physical array, then – in case of a line T/R array – the length of the synthetic one is:

$$L_{sa} = R_0\theta_3 \approx R_0\frac{\lambda_0}{L_r}$$

where R_0 is the minimum target range and L_r is the length of the physical array. The azimuthal resolution of the synthetic array is thus constant and equal to half the physical one:

$$\delta_{sa} = \frac{L_r}{2}$$

This simple expression is very important and shows a major difference between standard and synthetic aperture: for a standard aperture the longer the physical array the better the resolution, whilst for a synthetic one the shorter the physical array the better the resolution.

In the SITAR project, synthetic processing was applied to both primary and secondary bandwidths. Two high-frequency chirps[1] were used to generate a low-frequency one. By filtering, the received signal will then be split into three frequency ranges: HF1, HF2 and LF. SAS processing was applied to all frequency ranges and led to three images (for each array) of the same spot with comparable resolutions.

[1] A chirp is a time-limited sinusoidal signal.

3

Multiple-aspect scattering (MAS) measurements

Ph. Blondel, N. G. Pace, N. Jayasundere, M. Cosci, J. M. Hovem and M. A. Larsen

3.1 INTRODUCTION

The SITAR project aims at developing acoustic methods and instrumentation for imaging waste barrels/containers of small dimensions buried in unconsolidated sea sediments. As explained in Chapter 1, the objective was pursued along two main research lines. The first (Chapter 2) investigated and tested a conceptually innovative sonar system based on the principle of bottom-penetrating parametric sonar, but with the imaging capabilities of traditional side-scan sonar. The second research line focuses on the study and experimental verification of multiple-aspect acoustic sub-bottom scattering measurement techniques, and related automatic data processing strategies. This approach is based on the experimental observation that most acoustic energy is scattered by targets away from the backscatter direction, and that its distribution will strongly vary depending on the type of object and surrounding seafloor (Figure 3.1, see colour section).

Emergent technologies such as autonomous underwater vehicles (AUVs), and common instruments like remotely operated vehicles (ROVs), enable the acquisition of multiple-aspect scattering measurements in the exact configurations desired. The geometry of the imaging/receiving system can be adapted in real time to optimize the acquisition process, and make it possible to acquire multiple views of the buried objects as a result of their 3D reconstruction and characterization. This approach was selected for the SITAR sea trials, and modellled both theoretically (Larsen and Hovem, 2003) and in scaled experiments (Jayasundere and Blondel, 2004).

3.2 IMAGING STRATEGY AND SAMPLING REQUIREMENTS

Theoretical studies (e.g., Williams and Jackson, 1996) and experiments in the laboratory (e.g., Blondel *et al.*, 2002) or at sea (e.g., Williams and Jackson, 1998) have shown

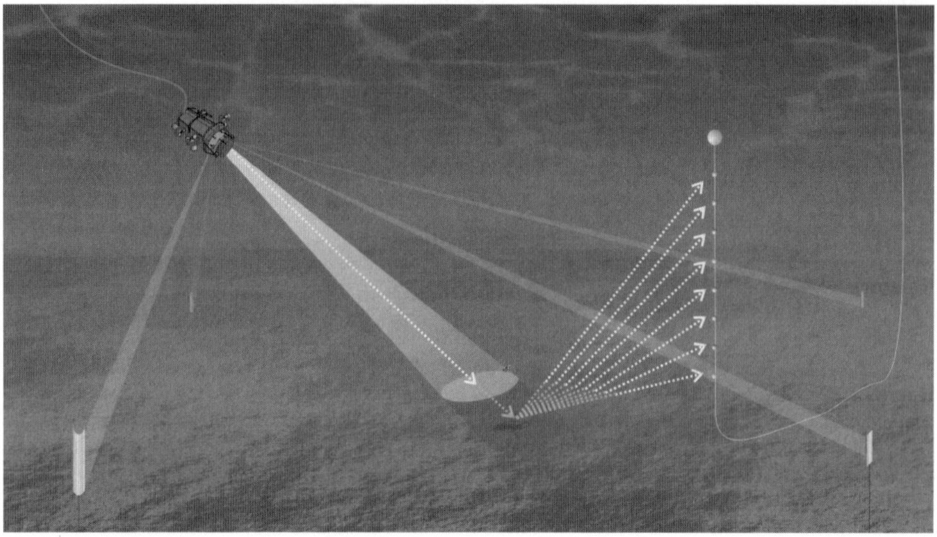

Figure 3.2. Multiple-aspect scattering measurements can be obtained with an acoustic transmitter imaging a patch of seabed and several receivers (for example, on a hydrophone chain) acquiring measurements at different scattering angles. This was the set-up adopted during the SITAR tank experiments and during the SITAR sea trials (Section 3.3). Transmitters and receivers need to be accurately positioned, in this case with a long-baseline transponder net.

the importance of using very different geometries of acquisition. As shown in Figure 3.2, multiple-aspect scattering measurements can be obtained using a narrow-beam acoustic projector, ensonifying a portion of the seafloor (which might contain one or several targets). The scattered acoustic energy is measured for several scattering angles – for example, by using receivers spaced on a chain. Different incidence angles will be achievable by moving the acoustic projector on a line above the target. And different bistatic angles can be achieved by circling specified patches of ground (for example, those in which targets have already been identified). Variations in the imaging geometry will have direct consequences on the scattering strengths measured (e.g., Blondel *et al.*, 2002; Canepa *et al.*, 2002). The positions and attitudes of the projector and receivers can be measured accurately using platform-based sensors and long-baseline transponder nets.

The theoretical studies done by NTNU (Larsen and Hovem, 2003) have constrained the sampling requirements and defined an experimental strategy for the accurate image reconstruction of a buried container with minimum dimensions of 0.1-m diameter and 0.7-m length, buried no deeper than 1 m. In this case, they assumed the acoustic projector would be the TOPAS parametric array, mounted on an ROV. Their recommendations can be summed up as:

- Measurements with different types of transmitted waveforms (chirp, Ricker) and different frequency ranges should be carried out. This is in order to exploit the advantages (resolution, bottom penetration, energy level) of each type.

- The positioning accuracy of the ROV should be good in order to have a reasonable estimate of the angle of incidence, and which part of the bottom is ensonified. Furthermore, the narrow beamwidth of the parametric source (as used in the SITAR sea trials) increases the demand for good ROV position accuracy. Sound velocity profiles should be acquired in order to calculate the exact travel path of sound from source to receivers.
- Sufficient coverage and sampling of the scattered field is critical for 3D image reconstruction of the target. The ROV will fly 15–20 m over the bottom. The hydrophone chain will be positioned about 15 m from the known target. With eight hydrophones in the chain 2 m apart, a distance of about 5 m above bottom for the lowermost hydrophone should be chosen. Results from scaled tank experiments at the University of Bath (Jayasundere and Blondel, 2004) suggest that this configuration will cover a wide range of the scattering angles of interest. Measurements with the TOPAS source oriented so as to produce lots of different combinations of incidence angle and bistatic angle should then be performed. Incidence angles producing specular reflection arrivals at the hydrophones should be prioritized. Results from the scaled tank experiments suggest that repeated measurements from the source held at the same position increase the signal-to-noise ratio by averaging the received signals. The bottom should also be ensonified without a target to make geo-acoustic inversion possible.

In order to produce an acoustic 3D image to be analysed/processed in the decision support system of the GIS-based data integration software (see Chapter 6), it is necessary to associate a relative scattering intensity to each x–y–z-coordinate of the seabed volume illuminated by the acoustic beam. The scatterers are idealized as a collection of point reflectors in the investigated volume. The processing chain developed within SITAR has been the following: first, an adaptive deconvolution process (Dobbins *et al.*, 2003) is applied to extract the arrival time at the various hydrophones of the signals of each scatterer. Then, a localization procedure is started. Referring to the geometry of Figure 3.3, assuming we know the source/receiver position, the constant sound speed along the travel path – so that the travel path is composed of straight lines – and the travel time along the path as determined from the deconvolution process, the scattered coordinates are determined by geometrical triangulation using information on all the available receivers.

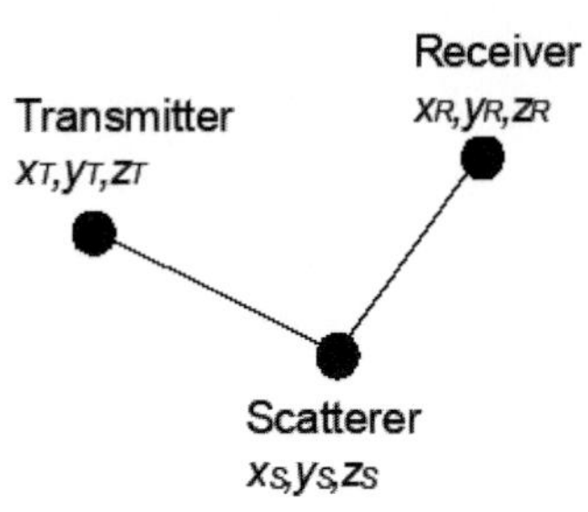

Figure 3.3. Propagation geometry for the localization process.

Having multiple receivers, with several sources of uncertainties, the localization problem is solved for each scatterer in a least-squares sense, by a downhill simplex method. An example of slices of a 3D acoustic image thus generated is given in Figure 3.4, where the acoustic image refers to a buried tyre-like object; the acoustic data were generated through simulation, for testing purposes.

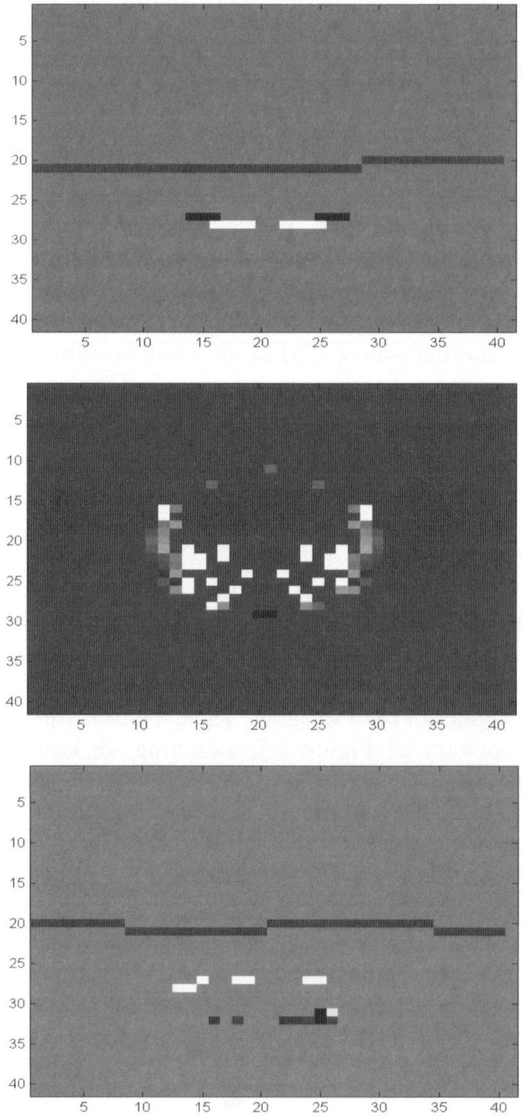

Figure 3.4. 3D acoustic image built with MAS simulated data; x–y–z-points (voxels) are associated to a relative scattering intensity level. From left to right, volume slices corresponding to planes parallel to the x–z, x–y, y–z planes, respectively.

3.3 MULTIPLE-ASPECT SCATTERING EXPERIMENTS

Very few sea experiments have been performed so far, and most of them have taken place in selected shallow-water sites on the continental shelf (e.g., Williams and Jackson, 1998; Blondel *et al.*, 2002; Canepa *et al.*, 2002). These experiments are fraught with difficulties, not all of which could be expected. One can, for example, list unexpected variations in the seabed, unexpected variations in the water column or unexpected effects (e.g., ship hull reflections, target burial during the time of the experiment). Because of their complexity, the results of most experiments are still being fully analysed. They have however revealed the complex interactions between targets and their environments. For this reason, we decided to first conduct scaled experiments in the laboratory, to assess optimal imaging geometries, test the processing and image reconstruction algorithms, and in general inform the sampling strategy that would be later adopted during the SITAR sea trials. Laboratory experiments are easier to perform, at least in theory: environmental variations can be controlled; the target emplacement is known with very high accuracy; target–background interactions can be directly measured and experimental uncertainties (e.g., transducer positions) are tightly controlled. Using the facilities of the University of Bath, we have conducted scaled experiments in a large water tank containing several sediment trays representative of continental margin seabeds (Figure 3.5). We used scaled targets comparable with the toxic waste containers found in many dumpsites at sea. Multiple targets (up to four) were imaged with a narrow acoustic beam at different incidence

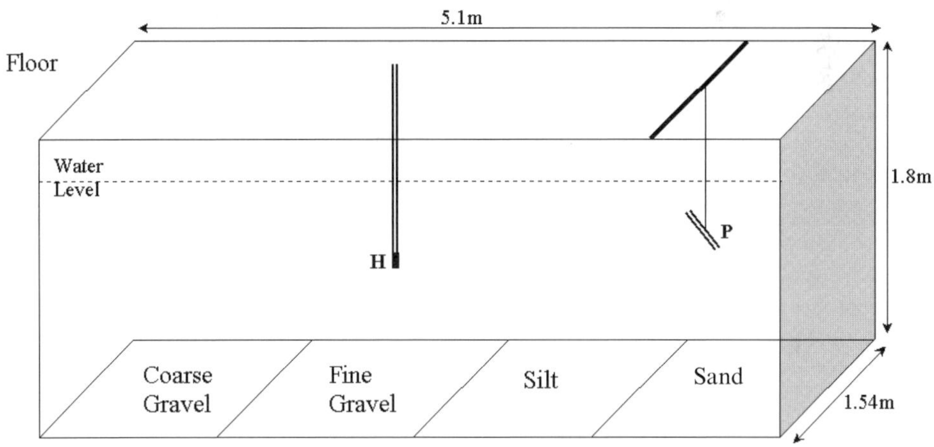

P – Projector, H – Hydrophone.

Figure 3.5. The University of Bath underground tank facilities. The walls are made of concrete, and the top of the tank is at floor level. The sediments in the trays are 14 cm thick on average. The water level can be varied. Both the acoustic projector and the hydrophone(s) can be positioned anywhere in the tank. The targets can be oriented in any way: proud, semi-buried or flush-buried.

angles. The receiving hydrophones were positioned to correspond to a vertical linear array some distance away from the targets. This set-up is a scaled-down version of the SITAR sea trials later performed in September/October 2003 in the Stockholm Archipelago, in a totally stable and controlled setting. The influence of environmental and experimental uncertainties had already been modellled (and measured) in Blondel *et al.* (2001, 2002). A scaling factor of approximately 10 : 1 was used in the design of the experiments. The targets were designed to match the numerical models used during the SITAR project (see Chapter 4) and the types of targets likely to be encountered at the trial site. The sediment types used in this study are silt (50-μm mean grain size) and gravel (5-mm mean grain size). The former are a scaled-down version of the soft muddy sediments, with a minute content of gas, expected at the sea trial site (as confirmed later). Karasalo and Skogqvist (2004) used results from previous inversions at nearby sites to obtain tentative values of 1.047 for the sound speed ratio and 1.1 for the density of the sediments, quite close to the values for the silt tray (1.024 and 1.204, respectively). The results from these experiments (signal processing and 3D reconstruction) are given in Sections 8.2 and 8.3.

3.4 CONCLUSIONS

The theoretical definition of the optimal sampling requirements has enabled us to identify key parameters (signal types, positioning accuracy) and a clear surveying strategy (line scans and rotation scans of particular targets, combination of imaging geometries). The scaled experiments complemented these with quantitative measurements, and enabled us to define a set of systematic and thorough quality-assurance procedures, which were used to control the quality of data acquisition during the sea trials. Some critical challenges could be thus addressed (e.g., signal losses, positioning problems), which would have gone unnoticed otherwise. Quantifying the ranges of scattering variations due to the environment, the seabed or the imaging geometries has also helped to ascertain which variations can be attributed to the target's condition (e.g., pristine/corroded, full/empty).

4

Acoustic models for scattering

I. Karasalo and P. Skogqvist

4.1 INTRODUCTION

The purpose of modelling activities is to provide computational techniques for the analysis of acoustic scattering that are both computationally efficient and numerically accurate. The acoustic models thus developed are to be employed in the analysis of the multiple-aspect scattering experimental data.

Two modellling methods are considered, a full-field boundary integral equation (BIE) method and a fast approximate Ray–Kirchhoff (RK) method. Both methods use a frequency-domain formulation of scattering from a 3D object in a layered, range-independent medium. Transient scattering is handled by Fourier synthesis of mono-frequency fields.

The complex acoustic pressure $p_s(\boldsymbol{r})$ at a point $\boldsymbol{r}$ outside a scattering body can be written:

$$p(\pi)(\boldsymbol{r}) = p_i(\boldsymbol{r}) + p_s(\boldsymbol{r}) \tag{4.1}$$

where $p_i(\boldsymbol{r})$ and $p_s(\boldsymbol{r})$ are the acoustic pressure of the incident field – that is, the field that would be excited without the scatterer present – and the scattered field, respectively. By Green's theorem, at a point $\boldsymbol{r}_0$ outside the surface S of the scatterer

$$p_s(\boldsymbol{r}_0) = \int_S [g(\boldsymbol{r}, \boldsymbol{r}_0)\nabla p(\boldsymbol{r}) - p(\boldsymbol{r})\nabla g(\boldsymbol{r}, \boldsymbol{r}_0)] \cdot \boldsymbol{n}(\boldsymbol{r}) \, dS(\boldsymbol{r}) \tag{4.2}$$

where $\boldsymbol{n}(\boldsymbol{r})$ is the unit outward pointing normal of S at $\boldsymbol{r}$ and $g(\boldsymbol{r}, \boldsymbol{r}_0)$ is the Green's function of the outside layered medium. Letting $\boldsymbol{r}_0 \to S$ in Equation (4.2) gives a boundary integral equation (BIE) for the unknown function $p(\boldsymbol{r})$ and its normal derivative on the surface $r \in S$

$$\tfrac{1}{2} p_s(\boldsymbol{r}_s) + \int_S \left(\frac{\partial g(\boldsymbol{r}, \boldsymbol{r}_s)}{\partial n} p(\boldsymbol{r}) - g(\boldsymbol{r}, \boldsymbol{r}_s) \frac{\partial p}{\partial n}(\boldsymbol{r}) \right) dS(\boldsymbol{r}) = p_i(\boldsymbol{r}_s) \tag{4.3}$$

For a rigid scatterer $\partial p/\partial n = 0$ on S and then $p(\boldsymbol{r})$ is determined on S by Equation (4.3), outside a discrete set of eigenfrequencies of an associated interior problem, as detailed in the remainder of this section. For acoustically penetrable scatterers both $p(\boldsymbol{r})$ and $\partial p(\boldsymbol{r})/\partial n$ are unknown, and Equation (4.3) must be complemented by an additional equation derived from the interior dynamics of the scatterer. For a homogeneous fluid body the interior dynamics is described by a BIE of the form (4.3) with zero right-hand side, and Green's function:

$$g(\boldsymbol{r}, \boldsymbol{r}_0) = -\frac{e^{ik|\boldsymbol{r}-\boldsymbol{r}_0|}}{4\pi|\boldsymbol{r}-\boldsymbol{r}_0|} \tag{4.4}$$

where $k = 2\pi/\lambda$ denotes the wavenumber in the interior fluid.

4.2 THE BIE METHOD

In the BIE code XFEM-S developed, Equation (4.3) and the similar interior equation are discretized by collocation, using a high-order B-spline basis for functions on S. The Green's function of the outside layered medium is computed by adaptive transform integration and exact finite elements (Karasalo, 1994). The singularity of the integrand of the BIE is eliminated by transforming to "tilted" coordinates, and the integrals are computed by high-order numerical quadrature.

Since the BIE (4.3) for both the exterior and interior problems is of the first kind in the normal derivative $\partial p/\partial n$ their straightforward use for numerical solution will lead to loss of accuracy or even failure due to poor conditioning. As is well known (Burton and Miller, 1971), equivalent well-conditioned forms of the BIE are obtained by linearly combining (4.3) with the BIE derived from the gradient of (4.2) in a similar way as (4.3). Numerical integration of the resulting "hypersingular" BIEs is handled by splitting the second derivative of the Green's function into terms that are alternatively singular and nonsingular at the B-spline knots. The contributions from the singular terms are then known analytically, and the regular terms are computed by the high-order quadrature schemes used for (4.3).

For scatterers with rotationally symmetric shape about the z-axis, the discretized BIE system is transformed into block diagonal form by discrete Fourier transform over the azimuthal B-spline index. The resulting diagonal blocks are of moderate size ($K \times K$ where $K = M + k_G - 1$) and may be solved efficiently by conventional LU decomposition. For other scatterers, such as the box studied here, an iterative method for linear general systems (Saad and Schultz, 1986), enhanced by preconditioning with a solver for a nearly rotationally symmetric case, is used.

4.3 MODEL OF SCATTERERS

Box-shaped scatterers have been considered in the study, as a realistic model of most dumped waste barrels. The numerical data used in the following have been taken in

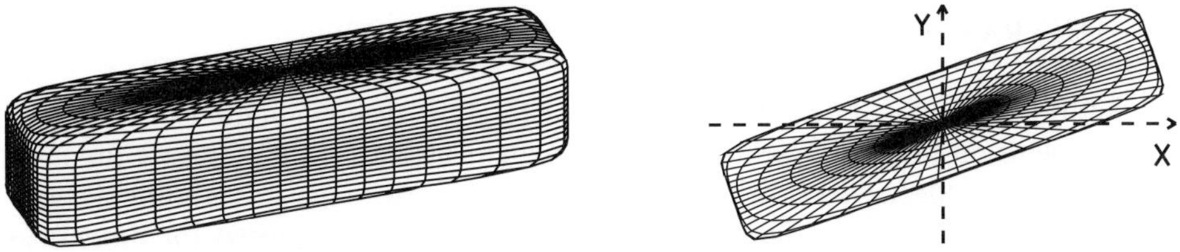

Figure 4.1. Computational model of the box: A homogeneous fluid body shaped as a super-ellipsoid with half-axes 65, 15 and 15 cm, exponent 10. Right: View from above. Yaw angle of box 19°.

order to resemble those of the dumped ammunition barrels encountered in the field trial. The scatterer is described by a smooth map of the spherical coordinates (θ, ϕ) onto the surface of a super-ellipsoid with half-axes 65, 15 and 15 cm as shown in Figure 4.1. The interior of the ellipsoid is homogeneous fluid with the following acoustic parameters: density 1630 kg/m^3, velocity 2680 m/s and absorption 0.7 dB/wavelength. The parameter values are representative of TNT explosive.

4.4 KIRCHHOFF'S APPROXIMATION

Kirchoff's approximation of the scattered field $p_s(\boldsymbol{r})$ is obtained by assuming a reflection coefficient $R(\boldsymbol{r})$ to be known, such that on the surface S, $p_s(\boldsymbol{r})$ is given by

$$p_s(\boldsymbol{r}) = R(\boldsymbol{r})p_i(r) \qquad \frac{\partial p_s(\boldsymbol{r})}{\partial n} = -R(\boldsymbol{r})\frac{\partial p_i(\boldsymbol{r})}{\partial n} \tag{4.5}$$

By inserting Equations (4.1) and (4.5) in (4.2)

$$p_s(\boldsymbol{r}_0) = \int_S [(1 - R(\boldsymbol{r}))g(\boldsymbol{r}, \boldsymbol{r}_0)\nabla p_i(\boldsymbol{r}) - (1 + R(\boldsymbol{r}))p_i(\boldsymbol{r})\nabla g(\boldsymbol{r}, \boldsymbol{r}_0)] \cdot \boldsymbol{n}(\boldsymbol{r})\, dS(\boldsymbol{r}) \tag{4.6}$$

$$p_s(\boldsymbol{r}_0) = -\int_S R(\boldsymbol{r})[g(\boldsymbol{r}, \boldsymbol{r}_0)\nabla p_i(\boldsymbol{r}) + p_i(\boldsymbol{r})\nabla g(\boldsymbol{r}, \boldsymbol{r}_0)] \cdot \boldsymbol{n}(\boldsymbol{r})\, dS(\boldsymbol{r}) \tag{4.7}$$

where (4.7) follows from (4.6) and Green's theorem is applied on $p_i(\boldsymbol{r})$.The reflection coefficient R is a function of the incidence angle $\theta_{inc} = \theta_{inc}(\boldsymbol{r})$ and possibly of frequency f, and it is defined as the reflection coefficient of a transversally homogeneous layered medium with structure equal to that of the scatterer locally at $\boldsymbol{r}$. The direction of incidence is defined as the direction of the intensity vector of the incident field I_i, and the reflection coefficient $R(\boldsymbol{r})$ is 0 at points $\boldsymbol{r}$ on the surface S where $I_i \cdot \boldsymbol{n}(r) > 0$ (non-insonified points).

4.5 THE RAY–KIRCHHOFF (RK) METHOD

Kirchhoff's approximation (4.7) is computed by numerical evaluation of a surface integral, with an integrand defined by the incident field $p_i(\boldsymbol{r})$, the Green's function $g(\boldsymbol{r}, \boldsymbol{r}_0)$ and the reflection coefficient R at the scatterer. Although this is computationally much less demanding than the BIE method, the work required for numerical evaluation of the transform integral for $p_i(\boldsymbol{r})$ and $g(\boldsymbol{r}, \boldsymbol{r}_0)$ in a layered medium may still make it impractical as a forward model for iterative parameter identification. A method without this drawback is obtained by replacing $p_i(\boldsymbol{r})$ and $g(\boldsymbol{r}, \boldsymbol{r}_0)$ in (4.7) by their ray-theory approximations (Skogqvist and Karasalo, 2003).

In the special case when $\boldsymbol{r}_1$ and $\boldsymbol{r}_2$ are located in adjacent homogeneous half-spaces, ray approximation is obtained in two steps. First, Fermat's principle is used to derive a fourth-degree polynomial equation for the position of the corner point of the two-segment eigenray from $\boldsymbol{r}_1$ to $\boldsymbol{r}_2$. The corner point is then found by solving the fourth-degree equation with a fast iteration-free algorithm. The ray-theory approximation of $g(\boldsymbol{r}_2, r_1)$ is then:

$$g(\boldsymbol{r}_2, \boldsymbol{r}_1) = -\frac{T(k)}{4\pi(r_1 r_2)^{1/2}} e^{i(k_1 R_1 + k_2 R_2)} \tag{4.8}$$

where k_1, k_2 are the wavenumbers in the two half-spaces and k is the horizontal wavenumber at propagation along the eigenray. R_1, R_2 are the lengths of the two eigenray segments. $T(k)$ is the transmission coefficient at the half-space interface:

$$T(k) = \frac{2\rho_2\gamma_1}{\rho_2\gamma_1 + \rho_1\gamma_2} \tag{4.9}$$

where ρ_j and $\gamma_j = (k_j^2 - k^2)^{1/2}$ are, respectively, the density and the vertical wavenumber in half-space j. The factor $1/(r_1 r_2)^{1/2}$ accounts for the change of sound pressure amplitude induced by the change of the ray tube cross-sectional area. Denoting the incidence angles of the ray segment with θ_1 and θ_2, then:

$$r_1 = R_1 + \frac{c_2}{c_1 R_2} \quad \text{and} \quad r_2 = \frac{\cos\theta_2}{\cos\theta_1 R_1} + \frac{c_2 \cos\theta_1}{c_1 \cos\theta_2 R_2} \tag{4.10}$$

The gradient of $g(\boldsymbol{r}_2, \boldsymbol{r}_1)$ with respect to $\boldsymbol{r}_2$ is required in (4.7) and is computed by differentiating (4.8) and the ray endpoint $\boldsymbol{r}_2$ with respect to launch angles and the arc-length, and then using the chain rule to convert to derivatives with respect to the components of $\boldsymbol{r}_2$. The integral (4.7) is formulated in tilted spherical coordinates θ', φ' with the north pole at the point of the insonified region where $I_i \cdot \boldsymbol{n}(\boldsymbol{r})$ is minimal. The boundary $\theta'_b(\varphi')$ of the insonified region $0 < \theta' < \theta'_b(\varphi'), 0 < \theta' < 2\pi$ is approximated by an interpolating spline, and the integral (1.8) is computed by high-order adaptive quadrature in θ' and the trapezoidal rule in φ'.

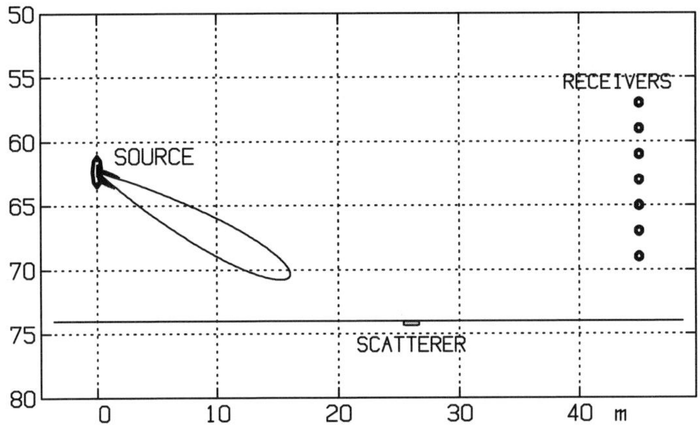

Figure 4.2. Experimental geometry of model case.

4.6 BIE VS. KIRCHHOFF AND RAY–KIRCHHOFF IN A MODEL CASE

In this section we present predictions of transient scattering with the BIE, the Kirchhoff and the Ray–Kirchhoff methods in a model case, with a buried scatterer and the source and a vertical receiver array configured as in Figure 4.2. The model case is similar to the multiple aspect scattering survey experimental configuration.

The scatterer is the box-shaped fluid body in Figure 4.1 buried 5 cm below the seafloor. The centre of the box is in the plane through the source and receivers, and the long and one short axis are horizontal. The long axis of the box is rotated by 19° out of the source–receiver plane as shown in the right half of Figure 4.1. The medium is composed of a homogeneous water half-space with sound speed $c = 1438$ m/s above a homogeneous fluid seabed half-space with parameters $c = 1400$ m/s, $\rho = 1110$ kg/m^3 and absorption $\beta = 0.01$ dB/wavelength. The source is modellled as a 9-m-long vertical array of 65 uniformly spaced point sources weighted and steered to emit a Ricker-like pulse with centre frequency 5 kHz within a side-lobe free beam directed towards the scatterer.

The top frames of Figure 4.3 show the transient scattered field $p_s(t)$ at the vertical array predicted by the RK (dotted) and the BIE methods (solid). The bottom frames show the RK (dotted) and the transform-integral Kirchhoff predictions in the same case. The dashed cross-trace curves in the left frames show the arrival times of the specularly reflected scattering component, as predicted by ray theory.

Both the Kirchhoff and the Ray–Kirchhoff method are seen to predict the arrival times and the duration of the pulses, as well as the detailed shape of their leading halves, with satisfactory accuracy. Also, the overall pulse shapes – the number of visible peaks, their signs and arrival times – are predicted correctly. The amplitudes of the peaks in the trailing half of the pulse predicted by the Kirchhoff methods are, however, overestimated by between 1% (receiver 7) and 55% (receiver 2).

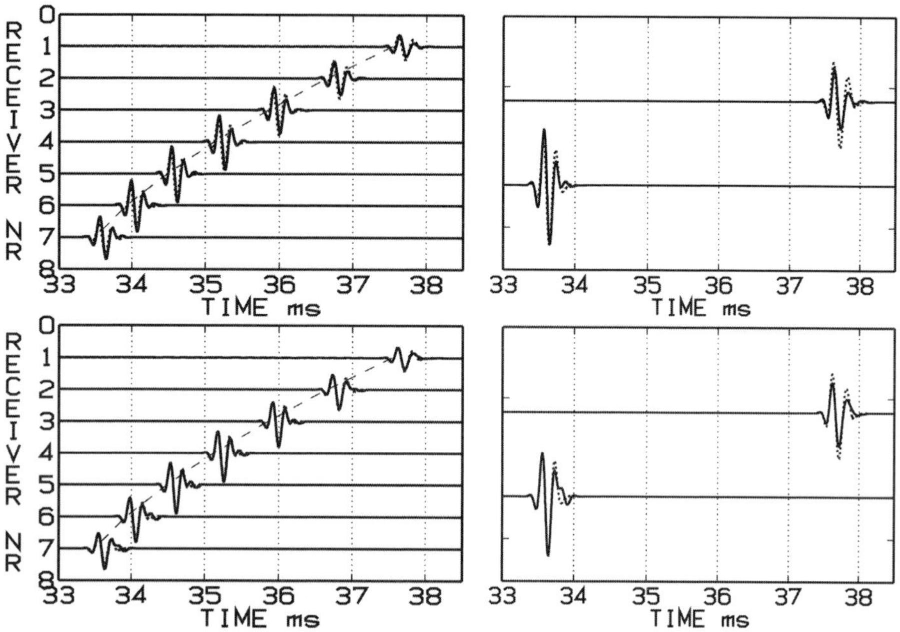

Figure 4.3. Model-predicted field scattered by a buried box-shaped fluid body. Above: BIE (solid), Ray–Kirchhoff (dotted). Below: Kirchhoff (solid), Ray–Kirchhoff (dotted). Left: All seven receivers. Right: Close-up of receivers 1 and 7.

As seen in the bottom frames the difference between the Kirchhoff and the RK results are significantly smaller than those between the RK and the BIE results. This indicates that the accuracy of the RK method is roughly the same as that of the transform-integral Kirchhoff method. The RK method is, however, much faster and therefore better suited as a forward model for iterative parameter identification.

5

Biotoxicity measurements: the nanoinjection technique

T. Hansson, G. Åkerman, U. Tjärnlund, K. Grunder, Y. Zebühr, H. Sundberg and L. Balk

5.1 INTRODUCTION

In the SITAR project the nanoinjection technique was used to inject sediment extracts into newly fertilised rainbow trout (*Oncorhynchus mykiss*) eggs, whereupon disorders in the developing embryos and larvae were recorded as a measure of potential toxicity.

The potential toxicity of munitions dumped on the seabed or buried beneath the sediment surface depends largely on the characteristics of the toxic substances. Earlier studies of the toxicity of dumping areas have focused mainly on the acute toxicity of hydrophilic (polar) substances and found very little toxicity, if any (Halsband, 1976; MHLCAKL, 1986; FLGWC, 1993; Russian Federation, 1993; HELCOM CHEMU, 1994; Henriksson *et al.*, 1996; Muribi, 1997; Muribi and Eriksson, 1997). A vast number of environmental pollutants are, however, lipophilic and many of these are not acutely toxic, but give rise to effects first after a long exposure period and/or long after the exposure. These environmental pollutants include well-known model pollutants, such as dichlorodiphenyltrichloroethane (DDT), polychlorinated biphenyls (PCBs), polybrominated biphenyls (PBBs), dioxins and dibenzofurans. Such persistent organic pollutants may bioaccumulate both by bioconcentration (directly from the water phase) and by biomagnification (via the food web). The parent substances are not necessarily toxic in themselves, but toxic metabolites may be formed during intracellular metabolism. These factors explain why adverse biological effects may not only be found at sites where these kinds of pollutants are released.

Exposure to lipophilic organic pollutants in the aquatic environment occurs in at least three ways: (i) by maternal transmission, (ii) by direct uptake from the water phase and (iii) by food consumption. With the nanoinjection technique used within SITAR all the above routes of exposure are taken into consideration. With nanoinjection, very roughly speaking, we refer to the introduction ("nanoinjection") of extracts from seabed samples of the study area in fertilized fish eggs. The growth disorders percentage observed in injected eggs from the study site with respect to a reference site gives a relative measure of accumulated potential toxicity at the site. The relevance of the method can be explained in the following way: surficial sediment is a relevant abiotic matrix for measurement of potential toxicity; extraction with toluene gives a relevant selection of lipophilic xenobiotics that are potentially bioaccumulated; nanoinjection of the extracts into newly fertilised rainbow trout eggs offers a relevant, sensitive biological system with characteristic vertebrate metabolism; and selected subcellular biomarker responses confirm relevant uptake over the cell membrane when a dose–response relationship is found after relevant chronic exposure.

As a positive control to the SITAR study area we also measured the potential toxicity along an 84-km-long pollution gradient through the Stockholm Archipelago (Sweden, Baltic Sea) with the city of Stockholm as a point source. The choice of control was motivated by documented toxic effects in adult feral perch (*Perca fluviatilis*) in this gradient (Balk *et al.*, 2003a, b). In addition to the biotoxicity measurement selected PCBs and polycyclic aromatic hydrocarbons (PAHs) were quantified in sediments from the stations of the Stockholm Archipelago gradient, including one site in the central part of the munitions dumping area.

5.2 SEDIMENT SAMPLING

The SITAR study area was the dumping area and its surroundings in the Möja Söderfjärd Bay and the reference area in the Möja Västerfjärd Bay, both located in the Stockholm Archipelago. Details on the archipelago and on the sediment sampling areas are given in Chapter 10, as well as the location of the individual sampling sites.

5.3 CHEMICALS

Toluene (*p.a.* grade), n-hexane (LiChrosolv®), silica gel (60, 0.063–0.200 mm), copper (*p.a.* grade, fine powder <63 μm), sulphuric acid (H_2SO_4, *p.a.* grade), potassium hydroxide (KOH, *p.a.* grade), cyclohexane (*p.a.* grade), dimethylformamide (DMF, *p.a.* grade) and sodium chloride (NaCl, *p.a.* grade) were purchased from Merck (Darmstadt, Germany). Triolein (T-7140, 99%), sucrose (*p.a.* grade), ethoxyresorufin (E-3763), resorufin (R-3257), ethoxycoumarin (E-1379), Tris-HCl (*p.a.* grade), reduced nicotinamide adenine dinucleotide phosphate (NADPH, N-0505, 98%), acetylthiocholine iodide (ACTC, A-5751, 98%), dithiobisnitrobenzoate (DTNB, D-8130), and Triton X-100 (T-9284) were purchased from Sigma (St. Louis, MO, USA). Hydroxycoumarin (H24003) was purchased from Aldrich (Steinheim, Germany). Benzo[a]pyrene (B[a]P) was purchased from Fluka Chemie GmbH (Buchs, Switzerland). Agarose gel was purchased from Pharmacia (Uppsala, Sweden). Native and ^{13}C-labelled polychlorinated biphenyl (PCB) standards as well as deuterium-labelled polycyclic aromatic hydrocarbon (PAH) standards were purchased from Cambridge Isotope Laboratories Inc. (Andover, MA, USA). Native PAH standards were purchased from well-established commercial sources.

5.4 EXTRACTION OF SEDIMENTS

The sediment samples and solvent blanks were extracted for 24 h in toluene using a Soxhlet apparatus connected with a Dean–Stark trap for water removal (Lamparski and Nestrick, 1980). After extraction the dry weights of the samples were determined gravimetrically. A 10 × 100-mm open silica column (10% water w/w) eluted with n-hexane was used for the removal of lipids and polar compounds. To remove sulphur the extracts were treated with a small amount of copper in an ultrasonic bath (4 × 15 min) and then left overnight at room temperature for completion of the reaction. An aliquot of each cleaned-up extract was put aside for quantification of PCBs and PAHs.

5.5 FRACTIONATION OF EXTRACTS FOR THE BIOTOXICITY MEASUREMENT

Exposure solutions for the biotoxicity measurement were prepared from the extracts. Half of each extract was retained as the "total extract". The other half was fractionated on a semi-preparative aminopropylsilica column (NH_2, µBondapak®; 7.8 × 300 mm, 10-µm particle size, 125-Å pore size; Waters, Milford, MA, USA) with n-hexane as the mobile phase in an automated high-performance liquid chromatography (HPLC) system (Zebühr *et al.*, 1993) with a UV–VIS detector operating at 254 nm. This straight-phase fractionation technique separated compounds predominantly according to their number of aromatic rings (Colmsjö *et al.*, 1987). The retention times for toluene and anthracene/phenanthrene were used to define two fractions, designated by their main components: (1) a dicyclic aromatic fraction, collected at a flow rate of 3 mL/min, and (2) a polycyclic aromatic fraction, collected upon back-flush of the column at a flow rate of 5 mL/min.

5.6 QUANTIFICATION OF PCBs AND PAHs

The aliquots put aside for quantification of PCBs and PAHs were spiked with seven ^{13}C-labelled mono- to tetra-*ortho* PCBs (IUPAC numbers 28, 52, 101, 105, 118, 138, 180) and five deuterium-labelled PAHs (D_{12}-phenanthrene, D_{10}-fluoranthene, D_{10}-pyrene, D_{12}-benzo[a]pyrene, D_{12}-benzo[ghi]perylene) before HPLC fractionation as described above. The dicyclic aromatic fraction was cleaned up on a triple-layer silica column (Pasteur pipette) deactivated with, from the top: H_2SO_4 (40% w/w), KOH (33% w/w) and water (10% w/w) (Zebühr, 1992), and with n-hexane as the mobile phase. The polycyclic aromatic fraction was cleaned up by liquid–liquid partitioning with DMF and cyclohexane, and elution through a 10 × 100-mm open silica column (10% water w/w) with n-hexane as the mobile phase (Näf *et al.*, 1992). After volume reduction under nitrogen gas, the dicyclic aromatic fraction and the polycyclic aromatic fraction were spiked with ^{13}C-labelled PCB 153 and D_{12}-chrysene, respectively, for calculation of internal standard recovery. The fractions were injected into a

Fisons GC8000 gas chromatograph (Thermo Electron Corp., Waltham, MA, USA) equipped with a PTE™-5 capillary column (0.25 mm × 30 m, 0.25-μm film thickness; Supelco, Bellefonte, PA, USA) and coupled to a Fisons MD800 mass spectrometer. The method for quantification was described by Axelman and Broman (1999). PCB concentrations were expressed as ng/kg dry sediment and PAH concentrations as μg/kg dry sediment. The accuracy of the PCB and PAH concentrations was generally 2 significant digits.

5.7 EXPOSURE OF FISH EMBRYOS AND LARVAE

Because of the large number of sampling sites the fish exposure experiments were performed on three separate occasions. The first experiment investigated the Möja Söderfjärd 1, 3, 4, 5 and 7 sites and the Möja Västerfjärd 1, 3 and 5 sites, and the second experiment investigated the Möja Söderfjärd 2, 6, 8 and 9 sites and the Möja Västerfjärd 2, 4 and 6 sites. The sites of the Stockholm Archipelago gradient (including Möja Söderfjärd 5) were investigated in the third experiment. Rainbow trout (*Oncorhynchus mykiss*) eggs and semen were purchased from Källefall-Baltak's fish farms (Brandstorp, Sweden). To maximize the homogeneity of the biological material only one family pair was used in each experiment. After artificial fertilization and water swelling, the eggs were placed in cup-shaped depressions in 1% agarose gel cast in square petri dishes (Falcon, Becton Dickinson Labware, Franklin Lake, NJ, USA) with 36 eggs/dish (Åkerman and Balk, 1995). The total extracts, dicyclic and polycyclic aromatic fractions, and solvent blanks were volume-reduced under a gentle stream of nitrogen gas and dissolved in the carrier substance, triolein, a major natural constituent in fish eggs, whereupon the remaining n-hexane was allowed to evaporate. The newly fertilized eggs were exposed to the triolein solutions using the nanoinjection technique (Åkerman and Balk, 1995; Walker *et al.*, 1996). The dose was equivalent to 12.5 g dry sediment per kg fertilized and water swollen egg. Corresponding doses of PCBs and PAHs per kg egg wet weight were calculated by multiplying the sediment-based concentrations by 0.0125. Three types of controls were used: non-injected controls, carrier controls (eggs injected solely with triolein) and solvent controls (eggs injected with solvent blanks). No effects of triolein or solvent could be detected. As a positive control, B[a]P dissolved in triolein was injected at doses of 0, 40, 200, 1000, 2500 and 5000 μg/kg egg wet weight. The triolein-based exposure solutions were drawn, using a vacuum suction pump, into aluminium silicate capillaries (Sutter Instrument Co., Novato, CA, USA) bevelled to sharp elliptical tips in a diamond grinder. The injections were performed under a stereomicroscope, with a one-dimensional hydraulic manipulator (Narishige, Tokyo, Japan) holding the glass capillaries. After penetration of the chorion and the vitelline membrane, the exposure solution was injected into the yolk by applying nitrogen gas pressure using a Pico-injector (Medical Systems Corp., Greenvale, NY, USA). Before injection, each capillary was individually calibrated by adjusting the nitrogen gas pressure and the time of injection. A volume of less than 1‰ of the egg volume was injected in each egg. Control and exposure groups were kept dark in identical individual flow-through systems (36 individuals/system) with an average temperature of 6.2 ± 0.2°C during the first experiment. During the second experiment the average temperature was 6.4 ± 0.2°C until the last 2 weeks, when the temperature was raised to 7.5 ± 0.2°C. During the third experiment the average temperature was 8.2 ± 0.2°C. Each flow-through system consisted of an aquarium containing 2 L charcoal-filtered tap water with an average flow of 17 mL/min, giving a turnover rate of 2 h. Embryonic and larval mortality was recorded every 1–3 days during development. Deformities, haemorrhages, heart sac oedema, yolk sac oedema, vertebral deformity and other morphological disorders were recorded just after hatch in experiments 1 and 2 and expressed as percent above non-injected and carrier control. Colour photographs of typical haemorrhages and yolk sac oedema have been published by Åkerman and Balk (1998). In the first experiment hatching took place 8.5 weeks after fertilization, and sampling was performed 5 weeks after hatching. In the second experiment hatching took place 8 weeks after fertilization, and sampling was performed 4 weeks after hatching. In the third experiment hatching took place 6.5 weeks after fertilization, and sampling was performed 4 weeks after hatching. The temperature and duration of the experiments differed for practical reasons, but were chosen so that each experiment was terminated after about 220 day-degrees (°C) post-hatch, corresponding to consumption of approximately $\frac{2}{3}$ of the yolk. The larvae were killed by severing the spinal column.

5.8 SAMPLING OF FISH LARVAE

The larvae were submerged in ice-cold saline under a stereomicroscope during measurement of the length and dissection of the liver. In the first and second experiment eight livers were pooled and homogenized at 0°C in 800 μL 0.25 M sucrose using a 0.5-mL Potter–Elvehjem homogenizer (size 18) with six up and down strokes at 400 r.p.m. Because of the limited size of the Potter–Elvehjem homogenizer, four livers at a time were homogenized in 400 μL 0.25 M sucrose, whereupon the two homogenates were pooled. The homogenates were immediately portioned into ice-cold cryo-tubes and submerged in liquid nitrogen. In the third experiment the same procedure was used, but three livers were pooled and homogenized in 300 μL 0.25 M sucrose. Hence, eight or three larvae corresponded to $n = 1$ for the calculation of enzyme activities. In the first and second experiments 40 livers were dissected from each exposure group and control group. In the third experiment 15 livers were dissected from each group. Hence $n = 5$ for the calculation of enzyme activities in all three experiments. Directly after liver dissection the muscles of the larvae were put into cryo-tubes and submerged in liquid nitrogen for determination of acetylcholinesterase (AChE) activity. Length, as a measure of growth, was expressed as percent above non-injected and carrier control, with negative values denoting retarded growth compared with control.

5.9 ETHOXYRESORUFIN-*O*-DEETHYLASE (EROD) ACTIVITY

Cytochrome P-450 1A (CYP1A) activity was measured in the liver as EROD activity according to the method described by Prough *et al.* (1978). The total extract was analysed in all three experiments, and the dicyclic aromatic fraction and the polycyclic aromatic fraction were analysed in experiment 3. Ethoxyresorufin and resorufin were dissolved in 0.1 M Tris-HCl buffer, pH 7.8, and kept dark at 4°C. A fresh solution of NADPH in distilled water was made each day of measurement and kept on ice. Liver homogenates were rapidly thawed, one at a time, under running tap water (30°C) using a Vortex® shaker to prevent temperature rise above 0°C in the sample. The thawed homogenates were put on ice until measurement was performed, within a few minutes. Enzyme activities in the liver homogenates were measured at room temperature (22 ± 1°C) with a spectrofluorometer (Jasco FP-777, Japan Spectroscopic Co. Ltd., Tokyo, Japan) in glass cells with a final volume of 2 mL. Emission intensities were recorded over time. The excitation wavelength was 530 nm and the emission wavelength was 583 nm. Background intensities in cells containing 10 to 30 μL liver homogenate and ethoxyresorufin to a start concentration of 5 μM were recorded for each sample before adding NADPH to a concentration of 50 μM to start the enzymatic reaction. All samples were measured in duplicate, and the linearity between protein content and enzyme activity was regularly checked. A standard curve was constructed from different concentrations of resorufin in standard buffer solution. The total enzyme activity of each liver pool (three or eight livers) was calculated from the emission intensities and expressed as pmol/liver/min. Average EROD activity in the liver homogenates from each exposure group was calculated and expressed as percent of carrier control. The absolute mean EROD activity ±95% confidence interval in the carrier control larvae, used for this calculation, was: 1.18 ± 0.52 pmol/liver/min (40 larvae, $n = 5$) in experiment 1, 1.30 ± 0.84 pmol/liver/min (40 larvae, $n = 5$) in experiment 2 and 0.67 ± 0.06 pmol/liver/min (30 larvae, $n = 10$) in experiment 3.

5.10 ETHOXYCOUMARIN-*O*-DEETHYLASE (ECOD) ACTIVITY

Cytochrome P-450 2B (CYP2B) activity was measured in the liver as ECOD activity in experiments 1 and 2 according to the method described by Prough *et al.* (1978). The procedure was the same as for measurement of EROD activity, but with the following changes: ethoxycoumarin and hydroxycoumarin were used instead of ethoxyresorufin and resorufin, respectively. The pH of the 0.1-M Tris-HCl buffer was 7.6. The excitation wavelength was 365 nm and the emission wavelength was 452 nm. The cells contained 240 μL liver homogenate and ethoxycoumarin to a start concentration of 0.1 mM.

5.11 ACETYLCHOLINESTERASE (AChE) ACTIVITY

AChE activity was measured in the muscle in experiment 3 and at half of the sites in experiment 2. The method for measuring AChE activity was described by Ellman *et al.* (1961). It uses acetylthiocholine iodide (ACTC) as a specific substrate and is based on the increase of yellow colour (wavelength 412 nm) produced from thiocholine reacting with dithiobisnitrobenzoate (DTNB). Each sample consisted of muscle tissue from three individuals and weighed 0.15-0.20 g. The samples were mashed at 0°C in 0.1 M Tris-HCl buffer, pH 8.0, with 0.1% Triton X-100 (muscle:buffer ratio 1:15 w:v) for 15–20 s using an Ultra-Turrax dispersing tool (Janke & Kunkel GmbH & Co. KG–IKA-Labortechnik, Staufen, Germany). The slurries were centrifuged at 10,000 g and 4°C for 20 min, and the supernatant was used for the AChE activity measurements. Reaction rates were read with a spectrophotometer (Hitachi U-3200, Hitachi Ltd., Tokyo, Japan) at 412 nm and 25°C. A 20-μL aliquot of the supernatant was added to 905 μL 0.1 M Tris-HCl buffer, pH 8.0, and after addition of 50 μL DTNB (0.01 M in 0.1 M Tris-HCl buffer, pH 8.0) the background reaction was recorded. The enzymatic reaction was started with 25 μL 0.1 M ACTC dissolved in distilled water. All samples were measured in duplicate, and the linearity between protein content and enzyme activity was regularly checked. The enzyme activity was expressed as nmol ACTC hydrolysed per minute per g larval muscle wet weight. The average AChE activity of five samples from each exposure group was calculated and expressed as percent of non-injected, carrier and solvent control in experiment 2 and of carrier control in experiment 3. (The choice of controls depended on the number of specimens and replicates available.)

5.12 NORMALIZATION OF EROD VALUES

To facilitate direct comparison between the three experiments, the EROD values were normalized. Differences in the background and response levels of EROD activity may be caused by several types of natural variation, such as larval size (and accordingly liver size), the exposure history of the mother and perhaps genetic differences in the basic level of CYP1A expression between batches of fertilized eggs. A possible cause for the lower response (when expressed as percent of control) in experiments 2 and 3 was that the larvae were smaller than those in experiment 1. Mean length $\pm$ 95% confidence interval for the non-injected and carrier control larvae was 26.6 $\pm$ 0.2 mm ($n = 80$) in experiment 1, and 23.3 $\pm$ 0.4 mm ($n = 80$) and 24.0 $\pm$ 0.2 mm ($n = 45$) in experiments 2 and 3. Since the larvae within each experiment came from the same family pair, this difference could be extrapolated to the rest of the material and may very well have resulted in smaller livers, and thus lower activity per liver in experiments 2 and 3.

The EROD values were normalized with respect to the response to B[a]P, which was measured as a positive control in each experiment. The normalization was performed after transformation of the values to percent of control. Since the response of the larvae in experiments 2 and 3 was low compared with other studies (see, for example, Balk *et al.*, 1995; Åkerman *et al.*, 2002; Sundberg *et al.*, 2003; Hansson *et al.*, 2004; Sundberg *et al.*, 2005a, b), the responses in experiments 2 and 3 were normalized in the following way to match the responses in experiment 1: linear regression was used to obtain a linear relationship between B[a]P dose and EROD response in experiment 1, and this was compared with the corresponding relationships in experiments 2 and 3. In experiment 2 the EROD response to B[a]P exposure was lower than in experiment 1 in the dose range 200 to 2500 μg/kg egg wet weight. These doses were associated with EROD responses between 200 and 600% of control – that is, the range that included the EROD responses to the extracts tested in experiment 2. For that reason doses up to 2500 μg/kg egg were used to obtain the linear relationship between B[a]P dose and EROD response in experiment 2. In experiment 3 the EROD responses to B[a]P exposure ranged from 100 to 700% of control and the EROD responses to the extracts tested in this experiment ranged from 100 to 1000% of control. Therefore, all tested B[a]P doses (0–5000 μg/kg egg wet weight) were used to obtain the linear relationship between B[a]P dose and EROD response in this experiment. The intercept was close to 100% of control in each of the three experiments. The slope in experiment 1 was 2.4 times higher than that in experiment 2 and 2.9 times higher than that in experiment 3, giving the following formulas for calculation of the normalized EROD values:

$$\text{EROD}_{\text{norm}} = 100 + 2.4 \cdot (\text{EROD} - 100) \quad \text{(Experiment 2)}$$

$$\text{EROD}_{\text{norm}} = 100 + 2.9 \cdot (\text{EROD} - 100) \quad \text{(Experiment 3)}$$

The EROD data for the exposure to the dicyclic and polycyclic aromatic fractions in experiment 3 were also normalized. Values lower than 100% of control were not normalized, since the formulas were not valid for such values. The EROD response to the extract from the Möja Söderfjärd 5 site was tested in both the first and the third experiment and was found to differ between these experiments. The difference was cancelled by the normalization, which was taken as a confirmation of its relevance.

6

Processing tools for acoustic 3D images

V. Murino, M. Palmese and A. Trucco

6.1 INTRODUCTION

To investigate man-made objects lying on the seafloor or embedded in the bottom, acoustic imaging techniques and subsequent data processing can be applied. In this chapter, we address the use of acoustic image processing to design a complete system for segmentation, reconstruction and final virtual rendering of underwater sub-bottom objects.

The analysis of volumetric acoustic images is a research field that is evolving: conventional approaches have not been yet devised. At present, different applications and different sensing configurations address distinct problems; thus, *ad hoc* solutions have been developed to address the SITAR objectives. Nevertheless, in general, there is the need to remove noise from raw data by performing preliminary processing. Image processing techniques to improve amplitude/intensity image quality are quite commonly utilized, mainly at speckle reduction and contrast enhancement (Alexandrou and De Moustier, 1988; Malinverno *et al.*, 1990; Sauter and Parson, 1994).

After image filtering, more structured post-processing methods can be applied, especially segmentation and reconstruction techniques for high-level tasks, like classification and object recognition. Depending on the type of original data to be analyzed, a variety of methodologies have been proposed to tackle the aforesaid issues; so it is not possible to identify an algorithm or even a specific methodology as "the standard" (Caiti *et al.*, 2003), even if statistical approaches can be recognized as the most commonly used in the literature, especially for segmentation problems. After the identification of image regions corresponding to objects present in the ensonified volume, methods able to extract certain features aimed at a better understanding of the acoustic image can be applied. Actually, such methods differ according to the various cases considered, and are often adapted to the specific image and to the kind of image representation (Murino and Trucco, 2000). Finally, techniques aimed at improving scene comprehension resulting in a 3D augmented/virtual reality representation can be very useful to assist a human operator to investigate the acoustic scene from multiple points of view (Murino and Trucco, 2000).

The processing chain presented in this chapter includes blocks devoted to noise reduction, statistical 3D segmentation, semi-automatic surface fitting, extraction of measurements and VRML (Virtual Reality Modeling Language) rendering. In addition, a multi-resolution data representation based on an octree approach can be applied at different steps of the chain, if necessary. In particular, the 3D segmentation method presented here is based on a volume growing approach, basically a 3D extension to traditional 2D region growing. Algorithm initialization consists in the choice of usually small volumes belonging to the different regions to be segmented. The volume growing operation is guided by a statistical approach based on optimal decision theory: at each step the algorithm estimates the parameters of a probability density function that adequately fits the current volume histogram. Concerning the surface fitting block, it is based on predefined geometric models: one of them is tentatively selected by the user after observation of the segmented object and is automatically or partially–manually adapted to the segmented data. Model invariance with respect to the pose and orientation of the object has been stressed and a metric to measure the goodness of the obtained results has been defined. The processing chain described in this chapter can be seen as a comprehensive tool that represents a novel and relevant achievement in the processing of underwater 3D images.

6.2 IMAGE PROCESSING CHAIN

The developed image analysis tool can be divided into four main phases (see Figure 6.1). Starting with the 3D raw image an optional filtering stage can be applied to reduce noise effects, thus preparing data for the segmentation task that allows one to partition the original image into regions representing the objects contained in the scene. The segmentation result is the input to the third phase, the surface reconstruction and parameter extraction stage, that based on geometrical models enables fitting a chosen shape to the cloud of points resulting from the previous step. Finally, the fourth stage is devoted to three-dimensional visualization of the obtained results. Moreover, an additional phase to generate a multi-resolution structure can be performed at any level of the processing chain, and hence it is described in Figure 6.1 as a transversal step to the typical scheme.

6.3 DATA REPRESENTATION

When a volume is organized as a 3D regular grid of voxels[1] it can be very efficient to build a multi-level structure. The octree is a multi-resolution pyramid for representation of three-dimensional spaces (Rodriguez *et al.*, 2000) where each level corresponds

[1] A voxel is a volume element, representing a value on a regular grid in 3D space (similar to the pixel, or picture element, in 2D space).

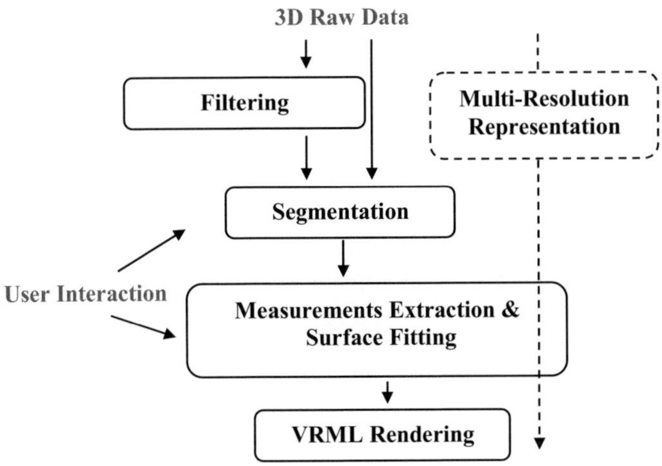

Figure 6.1. Image analysis processing scheme.

to a different resolution degree of the original image. The octree is built starting with the base of the pyramid – that is, from the finest-resolution level – and the resolution is iteratively reduced by spatially smoothing the available data. Considering the $(l-1)$-level, the corresponding image is divided in $2 \times 2 \times 2$ non-overlapping blocks; for each block the intensity levels of the non-empty voxels are averaged and will result in a single voxel in the upper level (l) (see Figure 6.2). By this procedure, it is possible to decrease data noise and to eliminate gaps in the data.

Hierarchical data representations make it possible to decide at which level of spatial resolution an algorithm can be applied. The same operation can work at the first level of the structure when the finest resolution is required, or it can be eased and sped up at a lower resolution level. A 3D image having a one-degree lower resolution in an octree scheme contains eight times fewer data, so it can be processed approximately eight times more quickly. Considering our processing scheme the octree structure can be applied at different levels of the chain depending on the current application:

- if gaps are present in raw data, the octree structure can be applied at the beginning of processing;

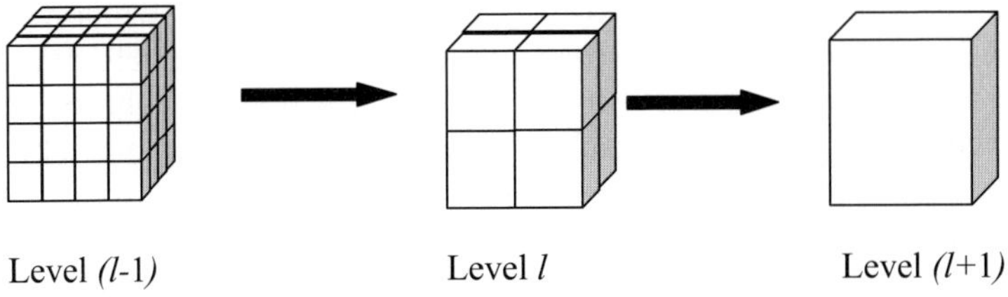

Figure 6.2. Octree structure generation procedure.

- when it is necessary to lighten the segmentation algorithm or the 3D rendering task, the operations can be performed at a poorer resolution level in the octree pyramid.

6.4 FILTERING

In addition to the spatial smoothing made possible by octree data representation, a 3D recursive median filter has also been implemented, in order to reduce speckle noise, which is characteristic of acoustic data. The median filter has been performed by a sliding three-dimensional mask with size $N \times N \times N$. The side dimension N must be an odd number and can be decided by the user according to the specific data. Appropriate size (from $3 \times 3 \times 3$ to $7 \times 7 \times 7$) must be chosen balancing the tradeoff between noise removal and image blurring. Moreover, as the median filtering is computationally intensive (the number of operations grows exponentially with the mask size) and to avoid an excessive loss in the data it is possible to apply a recursive implementation of the 3D filter choosing a small value for N. In this way, regions in which the signal period is lower than one-half the mask width will be continually altered by each successive pass until the resultant period is greater than one-half the window width, minimizing signal suppression. Also, the number of iterations can be adapted depending on the specific data needs.

6.5 VOLUMETRIC SEGMENTATION: STATE OF THE ART

To analyse 3D acoustic images of the sub-bottom and extract salient characteristics of buried objects a segmentation process is a fundamental step. The literature on the segmentation of 3D acoustic sub-bottom images is very limited, and the segmentation of 3D images is generally still considered a challenging problem for computer analysis. Therefore, it is necessary to also take into account 3D segmentation methods proposed for different application fields (e.g., medical imaging). Analysis of the state of the art has emphasized that the existing algorithms are difficult to classify in separate categories. Nevertheless, it is possible to subdivide them according to the following approaches: deformable surface models (balloons, level sets, bubbles), octrees, Markov random fields and volume growing.

In the balloon model, after initialization of a surface within the object of interest (manual procedure), the deformation process is based on minimization of the energy of a surface (Bowden *et al.*, 1997). Also, in the level set method (Baillard *et al.*, 2001) user interaction is required for the initialization task, but here the deformation surface is embedded at the zero level of a hyper-surface, which is the solution of a partial differential equation. With this formulation fast numerical methods can be applied to solve the problem. The issue of manual initialization can be resolved by the use of bubbles (Tek and Kimia, 1997). In the deformation process (reaction–diffusion process), after the random initialization of bubbles in the volumetric image, the bubbles grow, shrink, merge, split and disappear. For 3D acoustic images of the sub-bottom a fast 3D segmentation method based on octrees was proposed (Rodriguez *et al.*, 2000). The method is highly robust to noise and gaps in the data by performing the data processing in a multi-resolution pyramid – an octree. The tree is built by spatially smoothing the available data. This reduces noise and the number of gaps facilitating classification at a low spatial resolution. The main drawback of this method is the low spatial resolution. A way to incorporate spatial correlation and *a priori* information in the segmentation process is by use of a Markov random field (MRF) (Choi *et al.*, 1997). The MRF itself is a *conditional* probability model, where the probability of a given voxel status depends on its neighbourhood; the segmentation procedure is formulated as a maximum *a posteriori* estimation problem. Finally, a volume growing approach, 3D extension of 2D region growing, can be applied to segment a 3D image (Umesh and Chaudhuri, 2000). The segmentation task starts with a voxel or a set of voxels which belongs to the object of interest, examines its neighbourhood and decides for each voxel whether it belongs to the same object or not, according to a given criterion (stochastic approach – that is, MRF – watershed segmentation, dilation–erosion process).

6.6 THE VOLUME GROWING APPROACH

3D segmentation is the basic activity, while the pre-processing operations (e.g., filtering and enhancement) are aimed at preparing data to obtain good segmentation results, and can be arranged after definition of the 3D segmentation strategy. Image segmentation consists in identifying, inside a volumetric image, regions that are considered homogeneous in terms of some criterion. In our case segmentation is achieved by a "seeded" volume growing procedure. The choice of this method has been motivated by the fact that the volume growing approach combines different information about the image voxels: it considers both the spatial connectivity between voxels belonging to the same region and the similarity of the voxel intensity levels. This procedure needs to be initialized by a human operator, whose task is to select a (usually small) number of voxels belonging to different classes present in the image, according to the operator's judgement. These initialization voxels are called "seeds" (Adams and Bischof, 1994). Starting with the seeds, at each step of the algorithm the regions grow by addition of connected voxels in accordance with a measure of distance. The output of the volume growing procedure is a set of labelled volumes, where a label indicates the membership of a voxel in a segmented object. It is worth highlighting that initialization of the developed algorithm has been simplified in order to set the user free from the hard task of choosing complex parameters. The selection of the seeds is fast and user-friendly: starting with a 2D slice usually placed in the middle of the original volume the operator has to identify different regions corresponding to the different classes to be segmented. This first step of the segmentation procedure is the only one that requires user interaction and in fact provides all the information needed by the algorithm to go on automatically until the end of processing.

Let us suppose that n seeds have been selected, each corresponding to a class H_i, $i = 1, \ldots, n$. At each step of the algorithm, new voxels are added to some of the classes H_i by the following procedure:

(*i*) the unallocated voxels that border at least one of the regions are examined and labelled according to a measure of distance;
(*ii*) if the considered voxel q has more than one neighbouring region, we have to decide to which region voxel q is to be added. We calculate the distances from all its neighbouring regions, and add q to the closest region;
(*iii*) the algorithm ends when all the voxels have been allocated to the classes or the classes do not grow any more.

Let us describe the segmentation iterative procedure in detail. Let Q be the set of all the unallocated voxels q which border at least one of the regions $H_1, \ldots, H_i, \ldots, H_n$ already classified after a generic number of iterations. Remember that at the beginning of the process the sets $H_1, \ldots, H_i, \ldots, H_n$ are equal to the seeds selected by the operator. Thus, the set Q is defined as follows:

$$Q = \left\{ q \left| \left(q \notin \bigcup_1^n H_i \right) \wedge (q \in N(a)), \forall a \in H = \bigcup_1^n H_i \right. \right\} \tag{6.1}$$

where $N(a)$ is the set of immediate neighbours of voxel a. In particular, our choice for the set $N(a)$ is composed of the six voxels that are connected to voxel a.

At each step of the algorithm all the voxels of the Q set are labelled according to a measure of distance. If voxel q is adjacent to a different H_i it will be assigned to the region that satisfies the aggregation criterion. If the voxel is connected to only one region but such a region does not satisfy the condition of association, it will be temporarily labelled and included in a set of uncertain voxels; at the end of the iterative procedure this set will be reconsidered on the basis of the updated measure of distance. It is emphasized that a voxel is associated to a region if and only if it is connected to that region, as we have assumed connected components are to be segmented. At the end of each iteration the parameters of the aggregation condition are updated on the basis of the new voxels added to the classes. The algorithm is repeated until all the voxels have been allocated to classes or the classes do not grow any more. The segmentation algorithm can be outlined as shown in Figures 6.3, 6.4 and 6.5 where the schemes of the initialization step, the recursive phase and the final processing are respectively presented.

6.7 MEAN INTENSITY AGGREGATION CRITERION

The choice of aggregation criterion is critical to the success of the segmentation task. Two different criteria have been used in the segmentation procedure: the first is based on the intensity mean value of current regions; the second is based on the fitting of

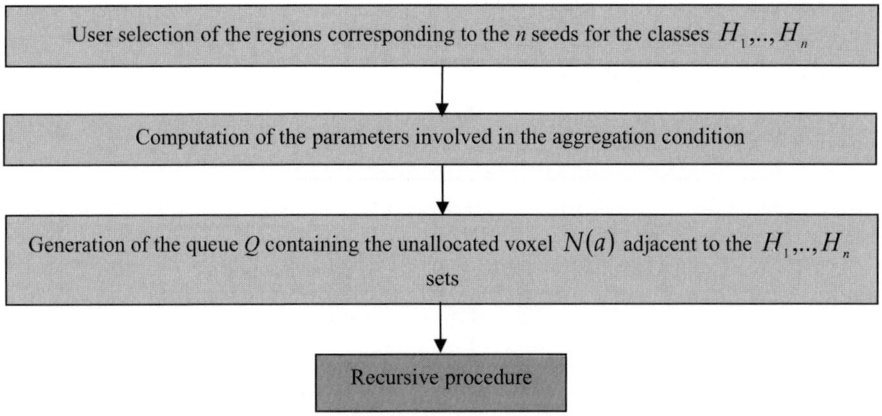

Figure 6.3. Scheme of the initialization step of the segmentation algorithm.

current volume histograms to an adequate probability density function (PDF). In the first case, the voxel is considered homogeneous to the region with the closest intensity mean value in terms of a measure of distance. The distance d is a simple quantity that shows how far the intensity (i.e., the scattering strength) of the considered voxel q is from the intensity mean value of the current region; it is defined as:

$$d_i(q) = \left| g(q) - \frac{\int_{H_i} g(y)\,dy}{\int_{H_i} dy} \right| \tag{6.2}$$

where $g(q)$ is the intensity level of voxel q. According to this criterion the algorithm chooses the kth region if the following expression is satisfied:

$$d_k(q) = \min_i \{d_i(q)\} \tag{6.3}$$

Actually, the effective assignment of voxel q to the chosen region must rely on the condition of spatial connectivity. This simple criterion is based on the hypothesis that assumes to work with image regions affected by Gaussian noise with zero mean and equal standard deviations. Even if for many images the assumption of equal variance noise is reasonable, we can modify distance d by taking into account the standard deviation SD of each growing region:

$$d_i(q) = \left| \frac{g(q) - \dfrac{\int_{H_i} g(y)\,dy}{\int_{H_i} dy}}{\underset{y \in H_i}{\mathrm{SD}}[g(y)]} \right| \tag{6.4}$$

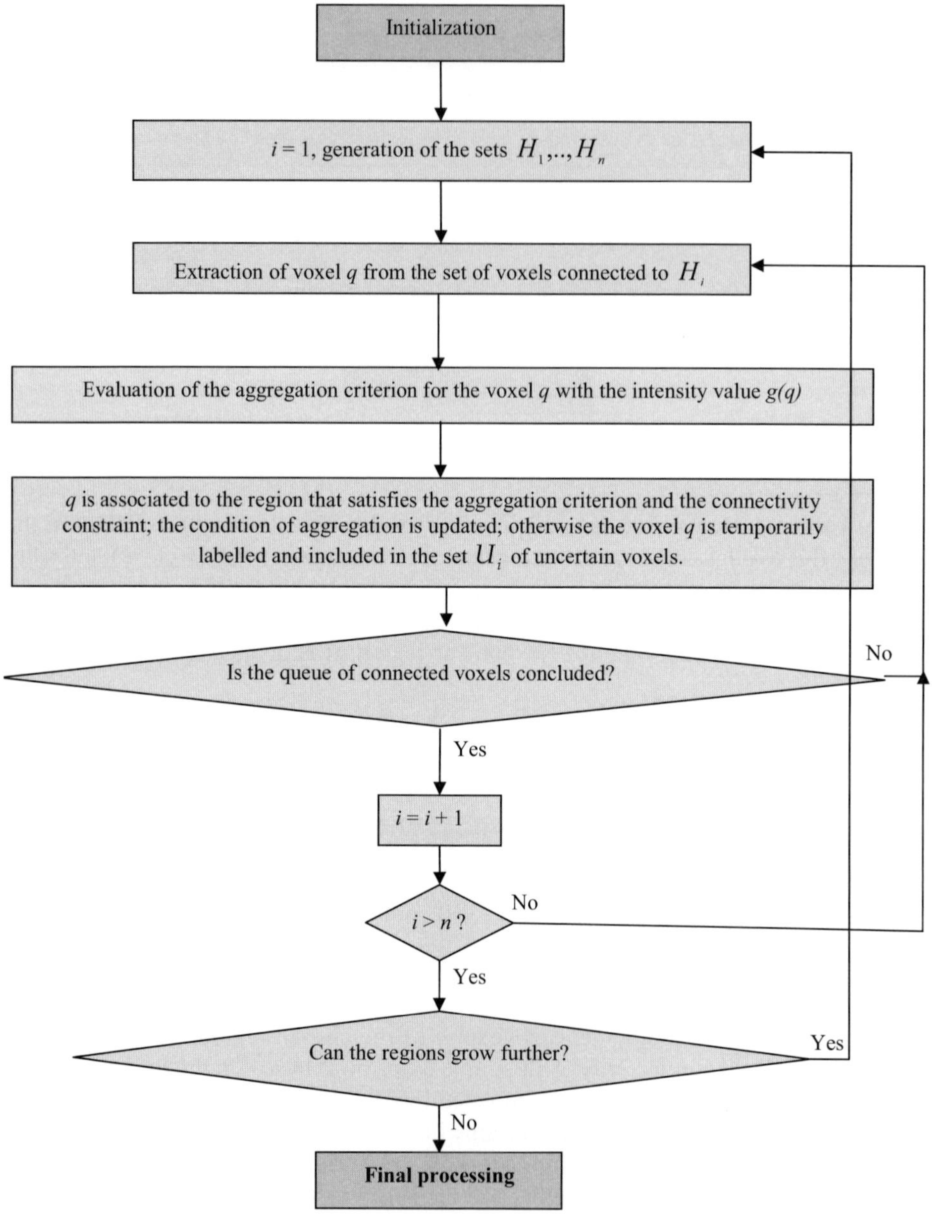

Figure 6.4. Scheme of the iterative phase of the segmentation algorithm.

However, this implementation of seeded volume growing is more computationally expensive without improving in an evident way the performances of the algorithm. Thus, when the assumption of Gaussian noise is not reasonable or *a priori* information is available a different formulation of the aggregation condition is necessary.

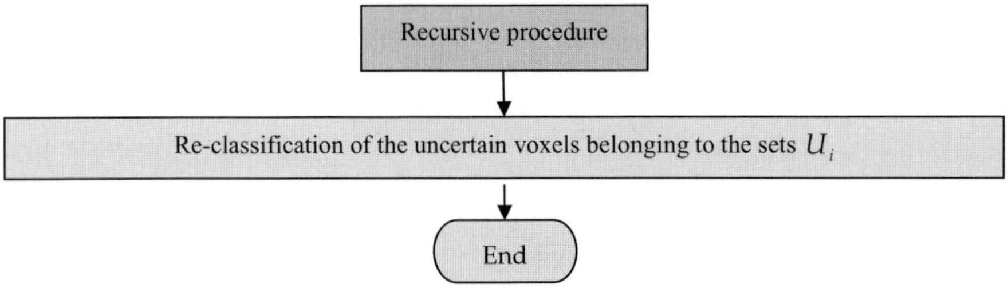

Figure 6.5. Scheme of the final step of the segmentation algorithm.

6.8 STATISTICAL AGGREGATION CRITERION

If *a priori* knowledge is available about the intensity distributions of the regions of the scene, it is possible to formulate a more accurate aggregation criterion taking into account second-order statistical information. This second solution (statistical volume growing) is based on the optimal decision theory: at each step of the algorithm and for each region, we calculate the current histogram and compute the parameters of the chosen density function (e.g., Gaussian, Rayleigh, Poisson density, etc.) that "best" fits the histogram. Each voxel connected to a region is added to that region if its intensity value satisfies the threshold condition based on the intersections of the current densities.

Voxel q, with an intensity level g and connected to a region H_i according to the estimated densities $p(g \mid H_i)$, will be assigned to that region if:

$$p(g \mid H_i) \geq p(g \mid H_k), \quad \forall k \neq i \tag{6.5}$$

This condition relies on the Bayesian approach based on the following hypothesis: if $\{H_1, \ldots, H_n\}$ are the available classes, the corresponding probability density functions $p(g \mid H_i)$ are known. Let us call $\{P_i\}$ the probabilities of the classes $\{H_i\}$ and R the Bayesian risk:

$$R = \sum_{i=1}^{n} \int_{Z_i} \left[\sum_{j=1}^{n} P_j \cdot C_{ij} \cdot p(g \mid H_j) \right] \cdot dg \tag{6.6}$$

where we have used the following notation: g is the voxel intensity level (feature) and is a stochastic variable, Z_i denotes the set of intensity levels belonging to the ith decision region, C_{ij} represents the cost we pay when voxel q is associated to H_i when it belongs to H_j. It is important to note that the probability $p(g \mid H_i)$ can assume the expression of any probability density function – for example, Gaussian or Rayleigh density – on the basis of statistical information about the specific component present in the 3D image. Moreover, we want to highlight that this criterion only takes into account the feature associated to each voxel; the spatial connectivity between the voxels belonging to the same region is considered through the volume growing process. In the case of non-sensitive cost classification the Bayes problem can be formulated as a minimization of the probability of error. Starting with Equation (6.6), defining the costs C_{ij} as follows:

$$\left.\begin{aligned} C_{ii} &= 0 \quad \forall i \\ C_{ij} &= 1 \quad \forall j \neq i, \forall i \end{aligned}\right\} \tag{6.7}$$

we obtain the consequent expression for the Bayesian risk:

$$r = \left\{ \sum_{i=1}^{n} \int_{Z_i} \left[\sum_{j=1, j\neq i}^{n} P_j \cdot p(g \mid H_j) \right] \cdot dg \right\} \tag{6.8}$$

In addition if we assume equal *a priori* P_i, where

$$P_i = \frac{1}{n} \qquad i = 1, \ldots, n \tag{6.9}$$

the quantity to be minimized becomes:

$$R = \frac{1}{n} \cdot \left\{ \sum_{i=1}^{n} \int_{Z_i} \left[\sum_{j=1, j\neq i}^{n} p(g \mid H_j) \right] \cdot dg \right\} \tag{6.10}$$

Let us denote by Z the set of all possible values of g

$$Z = \bigcup_{i=1}^{n} Z_i \tag{6.11}$$

and let us select the kth decision region Z_k:

$$Z_k = Z - \bigcup_{i=1, i\neq k}^{n} Z_i \tag{6.12}$$

By extracting the kth element from the external sum of Equation (6.10) and by including Equation (6.12) in Equation (6.10), the formulation for the risk is given by:

$$R = \frac{1}{n} \cdot \left\{ \sum_{i=1, i\neq k}^{n} \int_{Z_i} \left[\sum_{j=1, j\neq i}^{n} p(g \mid H_j) - \sum_{j=1, j\neq k}^{n} p(g \mid H_j) \right] \cdot dg + \int_{Z} \left[\sum_{j=1, j\neq k}^{n} p(g \mid H_j) \right] \cdot dg \right\} \tag{6.13}$$

where it should be recalled that a probability density function satisfies the condition of unitary area:

$$\int_{Z} p(g \mid H_i) \cdot dg = 1 \quad \forall i \tag{6.14}$$

Taking into account Equation (6.14), Equation (6.13) can be rewritten as:

$$R = \frac{1}{n} \cdot \left\{ \sum_{i=1, i\neq k}^{n} \int_{Z_i} [p(g \mid H_k) - p(g \mid H_i)] \cdot dg + (n-1) \right\} \tag{6.15}$$

Let us focus our attention on the integrals contained in Equation (6.15): the minimization of these terms coincides with the choice of decision regions that allow one to satisfy this condition:

$$\int_{Z_i} [p(g \mid H_k) - p(g \mid H_i)] \cdot dg < 0 \quad \forall k \neq i \tag{6.16}$$

namely, in all possible expressions of R in which the decision region Z_i appears, such a region will produce a negative contribution to the sum, thus reducing the Bayesian risk.

In conclusion, according to the Bayes criterion, we associate the voxel with intensity value g to the ith region if such a region satisfies condition (6.5): $p(g \mid H_i) \geq p(g \mid H_k)$, $\forall k \neq i$. It is worth observing that such a criterion follows the maximum likelihood (ML) rule, as we have assumed we are working with equal *a priori* probabilities. Actually, when this hypothesis is verified the ML method provides an optimal decision. Moreover, it is very favourable to adopt this method in terms of computational lightness. Furthermore, to lessen the computational load of the segmentation algorithm, it can be useful to update the parameters of the probability density functions of the regions only at the end of each iteration of the volume growing procedure. When there is lack of *a priori* information about the distributions of the intensity levels of the regions we are going to segment, the selection of a PDF with a single density shape will limit the performance of the segmentation approach. To overcome this problem it is possible to consider a density that allows us to control its kernel shape by selecting some parameters. The Weibull PDF is a family of densities of which the Rayleigh, Gaussian and exponential densities are special cases, corresponding to specific configurations of its parameters. The Weibull PDF of the intensity level g is given by:

$$f(g) = abg^{b-1} - e^{-ag^b} \quad g \geq 0, a, b > 0 \tag{6.17}$$

where a is a scale parameter and b is a shape parameter. Several methods, both graphical and analytical, are proposed in the literature to estimate the Weibull parameters (Al-Fawzan, 2005). Due to the high probability of error in using graphical methods and thanks to the availability of high-speed computers, we have preferred to use analytical methods –for example, method of moments (MOM), maximum likelihood estimator (MLE) and least squares method (LSM). The choice of the method depends on whether one needs a quick or an accurate estimate. In our case, to limit the computation time of the segmentation procedure, we have applied the last two methods, MLE and LSM, as – even though they yield less accurate results – they involve fewer calculations and require less computation time than MOM. In particular, LSM provides satisfactory estimates in a very short computation time. Examples of estimates are shown in Figure 6.6: the solid curve represents the histogram of the region we are considering; the dotted and the dashed lines indicate the Weibull PDFs estimated with LSM and MLE, respectively. It is possible to note that MLE is more accurate in fitting the original histogram (in the last part of the curve there is a total overlapping of the histogram with the estimated curve); nevertheless, LSM provides a good approximation in a shorter time without compromising the segmentation results.

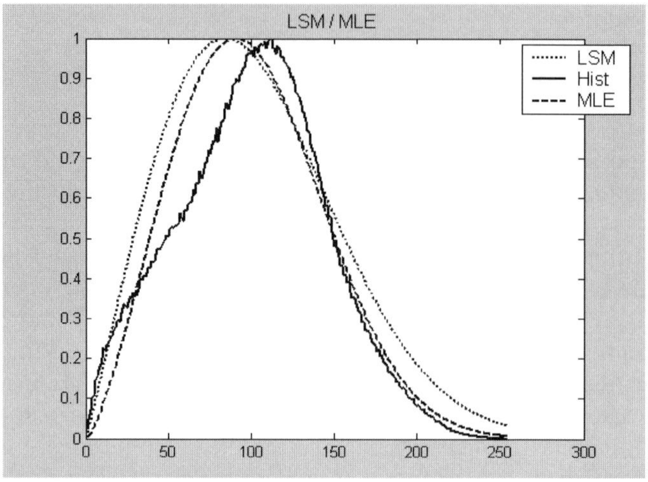

Figure 6.6. Examples of estimates of the Weibull PDF parameters by LSM and MLE.

6.9 PARAMETER EXTRACTION

Analysis of the segmentation results has the objective of extracting parameters useful to identify the spatial pose and the dimensions of objects eventually present in the scene of interest. To estimate such features a technique able to work with a wide assortment of object geometrical configurations has been developed.

The input to this step is a point distribution coming from the segmentation stage representing the natural or artificial objects we want to analyse. Generally, segmented data do not include the object in its wholeness, thus the algorithm must face the availability of partial data. The set of parameters we want to estimate concerns the pose of the object in space; namely, starting with an inertial coordinate system (ICS), defined by the vectors $\mathbf{v}_1, \mathbf{v}_2, \mathbf{v}_3$, we want to obtain the object coordinate system (OCS), defined by the vectors $\tilde{\mathbf{v}}_1, \tilde{\mathbf{v}}_2, \tilde{\mathbf{v}}_3$, fixed with respect to the object and taking into account eventual object symmetry. The first parameter we intend to find is the barycentre of the target. Unfortunately, the typical shortage of available data makes it impossible to give a correct measure of barycentre coordinates without an accurate morphological investigation. Generally, acoustical systems allow one to image only the surface of the objects within the scene, as acoustical energy is almost totally backscattered and only in small part is refracted inside the object. Therefore, initially we will limit ourselves to computing the barycentre coordinates $\mathbf{b}$ of the cloud of points arising from the segmentation process. The origin of the OCS is placed coincident with the distribution barycentre. To evaluate the inclination of the object, the rotation of OCS with respect to ICS has been expressed by means of the Euler angles (φ, θ, ψ). Starting with the inertial coordinate system the first rotation is by angle φ about the $\mathbf{v}_3$ axis (obtaining the vectors $\bar{\mathbf{v}}_1, \bar{\mathbf{v}}_2, \bar{\mathbf{v}}_3$), the second rotation is by angle $\theta \in [0, \pi]$ about the current $\bar{\mathbf{v}}_1$ axis (obtaining the vectors $\check{\mathbf{v}}_1, \check{\mathbf{v}}_2, \check{\mathbf{v}}_3$), and the third rotation is by an angle ψ about the current $\check{\mathbf{v}}_3$ axis (again), obtaining in this way the OCS (see Figure 6.7).

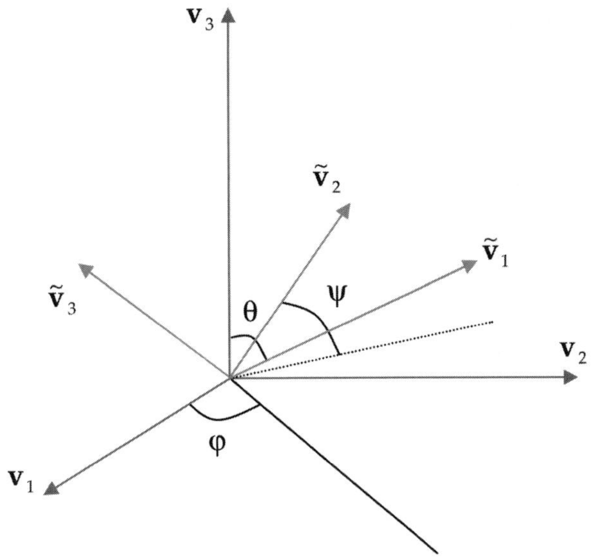

Figure 6.7. Adopted notation for the Euler angles, the ICS and the OCS.

Commonly, artificial objects can be assumed as rotational solids – namely, generated by a rotation of a surface about an axis. If we are in this condition we can neglect the ψ angle, as it does not provide any further information about the OCS. Therefore, for a rotational solid we can impose the following relation:

$$\psi = 0 \tag{6.18}$$

Let us assume we are working under this condition; to estimate the inclination of the object it is sufficient to compute the vector **t** defined as:

$$\mathbf{t} = [\theta \quad \varphi] \tag{6.19}$$

To obtain information on the orientation of a 3D distribution of points, it is possible to use a particular tensor associated to such a distribution, the inertial tensor. We assume that the point distribution has the same axes of symmetry of the object we want to reconstruct. Let us define the 3D distribution as the set of points in $\mathbf{R}^3$ with coordinates expressed in the ICS:

$$\{\mathbf{r}_i\} = \{x_i, y_i, z_i\}, \quad i = 1, \dots, N \tag{6.20}$$

Let us denote by m_i the mass corresponding to the ith point of the distribution. It is known that the barycentre **b** of a discrete distribution of N points is given by the following expression:

$$\mathbf{b} = \frac{\sum_{i=1}^{N} m_i \cdot \mathbf{r}_i}{M} \tag{6.21}$$

where M is the total mass of the distribution, equal to $M = \sum_{i=1}^{N} m_i$. As all the distribution points have been classified as belonging to the same object by the segmentation process, they are assigned the same weight in the barycentre formula. Thus, we set

$$m_i = 1 \quad \forall i \tag{6.22}$$

By including Equation (6.22) in Equation (6.21) it is possible to obtain the final expression for the distribution barycentre:

$$\mathbf{b} = \frac{\sum_{i=1}^{N} \mathbf{r}_i}{N} \tag{6.23}$$

As previously mentioned, the origin of the OCS has been set in $\mathbf{b}$. Let us consider a temporary coordinate system TCS, defined by vectors $\hat{\mathbf{v}}_1, \hat{\mathbf{v}}_2, \hat{\mathbf{v}}_3$, generated by a rigid translation of ICS in $\mathbf{b}$. Let us denote by $\hat{\mathbf{r}}_i$ the position vectors of the N points with respect to this new coordinate system and by $(\hat{x}_i, \hat{y}_i, \hat{z}_i)$ the respective coordinates.

After computing the new position vectors, $\hat{\mathbf{r}}_i = \mathbf{r}_i - \mathbf{b}$, we proceed by estimating the axes of symmetry of the distribution (see Figure 6.8). If $\rho(\hat{x}, \hat{y}, \hat{z})$ is a continuous distribution in $\mathbf{R}^3$, we define the inertial tensor $\mathbf{IT}$ of ρ as a 3×3 matrix given by the following formula:

$$\mathbf{IT} = \begin{bmatrix} \int \rho(\hat{x}, \hat{y}, \hat{z}) \cdot (\hat{y}^2 + \hat{z}^2) \cdot dV & -\int \rho(\hat{x}, \hat{y}, \hat{z}) \cdot \hat{x} \cdot \hat{y} \cdot dV & -\int \rho(\hat{x}, \hat{y}, \hat{z}) \cdot \hat{x} \cdot \hat{z} \cdot dV \\ -\int \rho(\hat{x}, \hat{y}, \hat{z}) \cdot \hat{x} \cdot \hat{y} \cdot dV & \int \rho(\hat{x}, \hat{y}, \hat{z}) \cdot (\hat{x}^2 + \hat{z}^2) \cdot dV & -\int \rho(\hat{x}, \hat{y}, \hat{z}) \cdot \hat{y} \cdot \hat{z} \cdot dV \\ -\int \rho(\hat{x}, \hat{y}, \hat{z}) \cdot \hat{x} \cdot \hat{z} \cdot dV & -\int \rho(\hat{x}, \hat{y}, \hat{z}) \cdot \hat{y} \cdot \hat{z} \cdot dV & \int \rho(\hat{x}, \hat{y}, \hat{z}) \cdot (\hat{x}^2 + \hat{y}^2) \cdot dV \end{bmatrix} \tag{6.24}$$

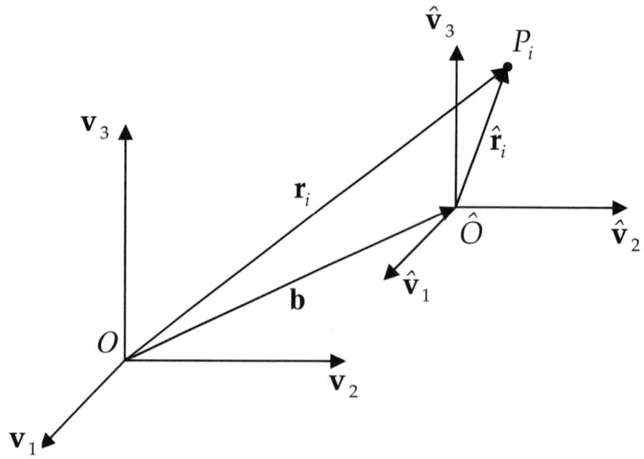

Figure 6.8. Rigid translation of the ICS in the distribution barycentre.

where dV is the infinitesimal volume element defined as:

$$dV = d\hat{x} \cdot d\hat{y} \cdot d\hat{z} \tag{6.25}$$

To obtain the same expression for a discrete distribution of N points we have to replace all the integrals with finite sums, and – taking into account the condition expressed by Equation (6.22) – we finally obtain:

$$\mathbf{IT} = \sum_{i=1}^{N} \begin{bmatrix} (\hat{y}_i^2 + \hat{z}_i^2) & -\hat{x}_i \cdot \hat{y}_i & -\hat{x}_i \cdot \hat{z}_i \\ -\hat{x}_i \cdot \hat{y}_i & (\hat{x}_i^2 + \hat{z}_i^2) & -\hat{y}_i \cdot \hat{z}_i \\ -\hat{x}_i \cdot \hat{z}_i & -\hat{y}_i \cdot \hat{z}_i & (\hat{x}_i^2 + \hat{y}_i^2) \end{bmatrix} \tag{6.26}$$

As **IT** is a real and symmetric matrix, it is always possible to perform a matrix diagonalization. This means that **IT** has non-degenerate eigenvalues e_1, e_2, e_3 and corresponding linearly independent eigenvectors. The eigenvalues of this tensor correspond to the rotation inertia of the distribution around the principal axes defined by the directions of the corresponding eigenvectors of **IT** (Giannitrapani *et al.*, 1999). In the case of a symmetric distribution, the eigenvectors of such a matrix, being the principal inertial axes, are symmetry axes too. Now, starting with the symmetry axes it is possible to get the Euler angles θ and φ necessary to describe the pose of the analysed object. We recall that angle ψ is neglected because of the hypothesis to work with a rotational object.

In Figure 6.9 the TCS denoted by vectors $\hat{\mathbf{v}}_1, \hat{\mathbf{v}}_2, \hat{\mathbf{v}}_3$ and the OCS denoted by $\tilde{\mathbf{v}}_1, \tilde{\mathbf{v}}_2, \tilde{\mathbf{v}}_3$ are shown; it is possible to note the θ and φ angles describing the rotation between the two systems.

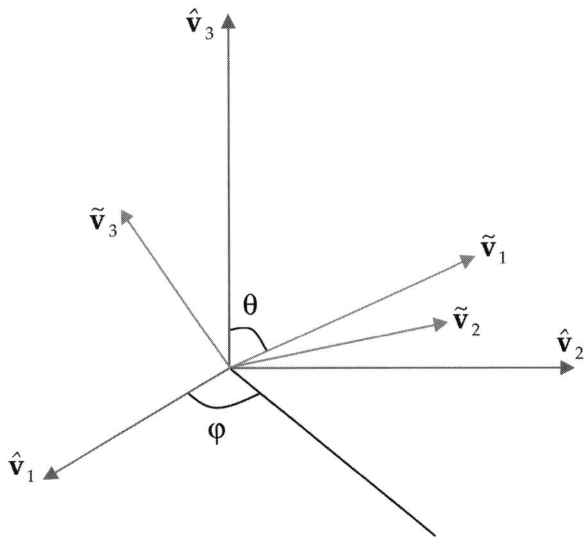

Figure 6.9. Rotation between the coordinate system denoted by vectors $\hat{\mathbf{v}}_1, \hat{\mathbf{v}}_2, \hat{\mathbf{v}}_3$ and the OCS expressed by the Euler angles.

To extract the coordinates of the OCS vectors with respect to the system TCS, it is necessary to compute the following scalar products:

$$\left.\begin{aligned}
\tilde{\mathbf{v}}_1 \cdot \hat{\mathbf{v}}_1 &= \sin(\theta)\cos(\varphi)\\
\tilde{\mathbf{v}}_1 \cdot \hat{\mathbf{v}}_2 &= \sin(\theta)\sin(\varphi)\\
\tilde{\mathbf{v}}_1 \cdot \hat{\mathbf{v}}_3 &= \cos(\theta)\\
\tilde{\mathbf{v}}_2 \cdot \hat{\mathbf{v}}_1 &= -\sin(\varphi)\\
\tilde{\mathbf{v}}_2 \cdot \hat{\mathbf{v}}_2 &= \cos(\varphi)\\
\tilde{\mathbf{v}}_2 \cdot \hat{\mathbf{v}}_3 &= 0\\
\tilde{\mathbf{v}}_3 \cdot \hat{\mathbf{v}}_1 &= -\cos(\theta)\cos(\varphi)\\
\tilde{\mathbf{v}}_3 \cdot \hat{\mathbf{v}}_2 &= -\cos(\theta)\sin(\varphi)\\
\tilde{\mathbf{v}}_3 \cdot \hat{\mathbf{v}}_3 &= \sin(\theta)
\end{aligned}\right\} \tag{6.27}$$

The coordinates of the three vectors $\tilde{\mathbf{v}}_1, \tilde{\mathbf{v}}_2, \tilde{\mathbf{v}}_3$ with respect to the TCS are equal to:

$$\left.\begin{aligned}
\tilde{\mathbf{v}}_1 &= \begin{bmatrix} \sin(\theta)\cdot\cos(\varphi)\\ \sin(\theta)\cdot\sin(\varphi)\\ \cos(\theta) \end{bmatrix}\\
\tilde{\mathbf{v}}_2 &= \begin{bmatrix} -\sin(\varphi)\\ \cos(\varphi)\\ 0 \end{bmatrix}\\
\tilde{\mathbf{v}}_3 &= \begin{bmatrix} -\cos(\theta)\cdot\cos(\varphi)\\ -\cos(\theta)\cdot\sin(\varphi)\\ \sin(\theta) \end{bmatrix}
\end{aligned}\right\} \tag{6.28}$$

Therefore, the rotation can be expressed by the following 3×3 matrix:

$$\mathbf{R} = \begin{bmatrix} \sin(\theta)\cdot\cos(\varphi) & -\sin(\varphi) & -\cos(\theta)\cdot\cos(\varphi)\\ \sin(\theta)\cdot\sin(\varphi) & \cos(\varphi) & -\cos(\theta)\cdot\sin(\varphi)\\ \cos(\theta) & 0 & \sin(\theta) \end{bmatrix} \tag{6.29}$$

It is now clear that the θ and φ angles can be easily computed starting with the components of the generic vector $\tilde{\mathbf{v}}_k$ – that is, we can consider the vector $\tilde{\mathbf{v}}_1$ and find the parameters performing the following procedure. Let us denote by v_{kj} the jth component of the kth vector $\tilde{\mathbf{v}}_k$. A possible expression for the θ angle is:

$$\theta = \arccos(v_{13}) \tag{6.30}$$

In this way we obtain an angle belonging to the interval $[0, \pi]$ according to the Euler convention. If $\sin(\theta) \neq 0$, the φ angle can be estimated starting with the φ' angle resulting from the expression:

$$\varphi' = \arccos\left[\frac{v_{11}}{\sin(\theta)}\right] \tag{6.31}$$

In this way we obtain again an angle belonging to the interval $[0, \pi]$, but to describe all the possible poses of an object it is necessary that

$$\varphi \in [0, 2\pi] \tag{6.32}$$

thus, the final φ is computed by taking into account the signs of the $\tilde{\mathbf{v}}_1$ components too:

$$\varphi = \begin{cases} \varphi' & \sin v_{12} \geq 0 \\ 2\pi - \varphi' & \sin v_{12} < 0 \end{cases} \tag{6.33}$$

Alternatively, if $\sin(\theta) = 0$, φ can be neglected as we are considering a rotational solid; thus, we can simply set $\varphi = 0$. In this way we have computed the angles θ and φ that identify univocally the direction of $\tilde{\mathbf{v}}_1$, which is a symmetry axis of the analysed object; we do not know yet which is the eigenvector that corresponds to such $\tilde{\mathbf{v}}_1$. As we are considering rotational solids, the computed angles make sense only if $\tilde{\mathbf{v}}_1$ is the rotation axis of the object. To identify the rotation axis among the three eigenvectors, user interaction is required. 3D data are projected onto the planes normal to the eigenvectors and the operator must choose the projection orthogonal to the axis of symmetry, identifying in such a projection an object with a circular (or partially circular) symmetry.

In conclusion, the developed method allows one to find the symmetry axes of an object starting with the corresponding point distribution. The proposed technique presents some limits when the point distribution resulting from the segmentation process represents only a fraction of the interested object, and thus is not characterized by the same properties of symmetry. This shortage of data produces two main effects:

- the barycentre position results up-shifted;
- the point distribution loses the symmetries of the original object, compromising computation of the θ angle.

To tackle this problem an *ad hoc* technique has been developed. The aim is to extract from the available data a portion that preserves the symmetry properties of the object. This operation works as a sort of further stage of segmentation to separate different regions of the same object and take out the region useful for the successive step. The first processing step concerns the extraction of the skeleton (Giannitrapani *et al.*, 1999; Murino and Giannitrapani, 1999). Before describing the applied procedure, a rough definition of skeleton is proposed – that is, a skeleton is a distribution of 3D points that:

(1) must be thinner than the original one;
(2) must be located in the neighbours of the median lines of the original point set;
(3) must have the same homotopy group; and
(4) have to be invariant to 3D rotation.

The third point requires, essentially, that the operations performed on the distribution to extract the skeleton preserve such topological information as the number of "holes" in the distribution and their relative position. In short, skeleton extraction is performed along the following lines. Let us define the 3D image as the ensemble of points in $\mathbf{R}^3$:

$$I = \{\mathbf{x}_i\} = \{x_i, y_i, z_i\}, \quad i = 1, \ldots, N \tag{6.34}$$

We also define for each point i and each R in $\mathbf{R}^+$ (ray of a sphere about the point i) the subset O_i^R of I and the 3D point $\mathbf{b}_i^R$ defined in the following way:

$$O_i^R \equiv \{\mathbf{x}_j \in I : |\mathbf{x}_i - \mathbf{x}_j| < R\} \tag{6.35}$$

$$\mathbf{b}_i^R \equiv \frac{\sum\limits_{\mathbf{x}_j \in O_i^R} \mathbf{x}_j}{\dim\{O_i^R\}} \tag{6.36}$$

where $\dim\{O_i^R\}$ is the cardinality of O_i^R (i.e., $\mathbf{b}_i^R$is the centroid of O_i^R). Let us define the *interior* λ_i^R for the point i in the following way:

$$\lambda_i^R \equiv \frac{R - |\mathbf{x}_i - \mathbf{b}_i^R|}{R} \tag{6.37}$$

One can easily verify that:

$$0 \leq \lambda_i^R \leq 1 \quad \forall i \tag{6.38}$$

The interior is a measure of how much a point is "inside" the object specified by I. If $\lambda_i^R \approx 1$, the point is inside a homogeneous zone of radius at least equal to R; in contrast, for $\lambda_i^R \approx 0$, the point is very near to a border of the three-dimensional distribution of points. Defining the following image transformation as:

$$I \to I^R = \{\mathbf{b}_i^R\} \tag{6.39}$$

and indicating by the symbol $I^{R,n}$ the iterative application of it for n times, our skeleton extraction is simply the construction of the image $I^{R,n}$ for a suitable choice of R and n. The overall effect of this transformation is to shift points on the border – that is, points with a low λ^R – toward the centre while leaving points well inside an object – that is, points with an almost unitary value of λ^R – unaltered. The iterative application of such a procedure tends to shift all the points of the distribution towards the skeleton. Such an algorithm may be seen from a physical point of view as a short-range interaction between physical points.

The choice of free parameter R is very important. If it is too small, with respect to the average mutual distances of the points, the distribution would likely collapse in a certain number of disconnected punctual regions, preventing extraction of the skeleton. If it is too large the value of λ may also be small for points well inside the object and again the skeleton will not be properly extracted. The parameter R plays a similar role to that of the dimension of the structuring element in classical mathematical morphology. It is straightforward to verify that the skeleton thus extracted is invariant for three-dimensional rotation (a condition that is never exactly satisfied for skeletons of a 2D bitmap image). Using *a priori* knowledge of the structure represented by the 3D image and of sensor resolution, it is possible to estimate proper values for R and n aimed at improving the extraction of a good skeleton. Thus, the choice of radius R has been formulated as:

$$R = \frac{ris}{2} \tag{6.40}$$

where *ris* is sonar system resolution. Also, the number of iterations n is relevant to obtain a good skeleton extraction. Unfortunately, it is difficult to find an objective measure of the goodness of a skeleton without *a priori* information about the real object shape. Thus, the stop criterion we have selected is based on local properties of the skeleton itself, in particular on the average number of points $\bar{n}_v$ which is contained in a neighbourhood. If the relative variation of $\bar{n}_v$ with respect to the previous value is little this means that points density is going to stabilize. In this situation it can be convenient to stop the iterative procedure. In particular, considering the kth iteration, the relative variation $\bar{n}_r(k)$ is given by the following formula:

$$\bar{n}_r(k) = \frac{\bar{n}_v(k) - \bar{n}_v(k-1)}{\bar{n}_v(k)} \tag{6.41}$$

where $\bar{n}_v(k)$ and $\bar{n}_v(k-1)$ are the average numbers of points in a neighbourhood during the kth and $(k-1)$th iteration, respectively. Such $\bar{n}_r(k)$ should be compared with an adequate threshold, chosen *ad hoc* for the specific case.

In Figure 6.10 the algorithm of skeleton extraction is outlined. In particular, we have highlighted the three successive phases:

(1) initialization (dark grey);
(2) recursive procedure (light grey); and
(3) final step (darker grey).

Because of the recursive nature of the algorithm and the possibility of working with a large number of distribution points, skeleton extraction could result in an onerous operation. To limit computational time, it is necessary to choose a threshold that minimizes the number of iterations and, at the same time, allows one to obtain a good contraction of the distribution. Figures 6.11 and 6.12 show a point distribution arising from the segmentation process and the relative extracted skeleton, respectively. The considered object is a cylinder, whose inclination is defined by non-null θ and φ; it is possible to note that the part of the cylinder imaged by the sonar system – and thus segmented – coincides with the upper surface of the whole target.

This is an example of partial data that requires the step of skeleton extraction to separate an object region needed to estimate the inclination of the whole object. In this specific case we can set apart the upper base of the cylinder and then perform the inertial tensor method to compute the Euler angles. We suppose that we have an estimation of the φ angle, provided by preliminary application of the inertial tensor

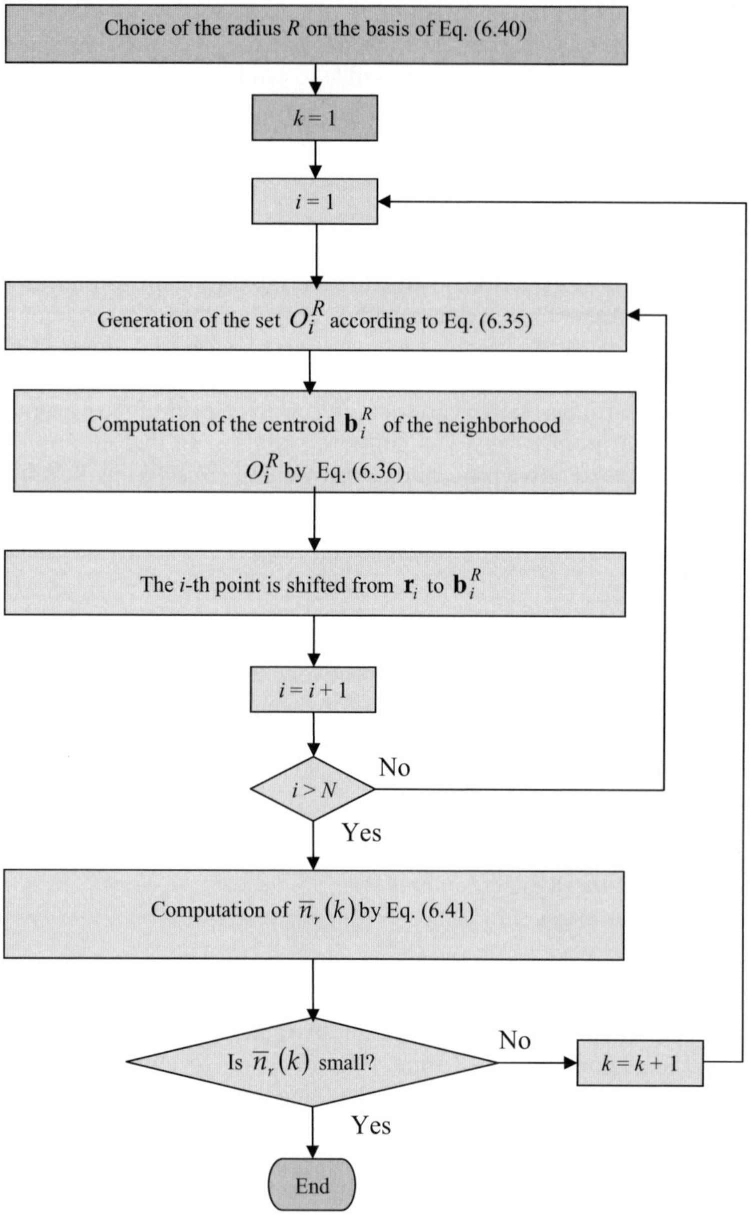

Figure 6.10. Scheme of the skeleton extraction procedure.

method. Our aim is to obtain an accurate measure of the θ angle, which provides the inclination to the normal to the upper cylinder base with respect to the vertical direction. The plane containing the upper base of the cylinder can be easily selected within the object skeleton: it can be identified as the plane that satisfies the following

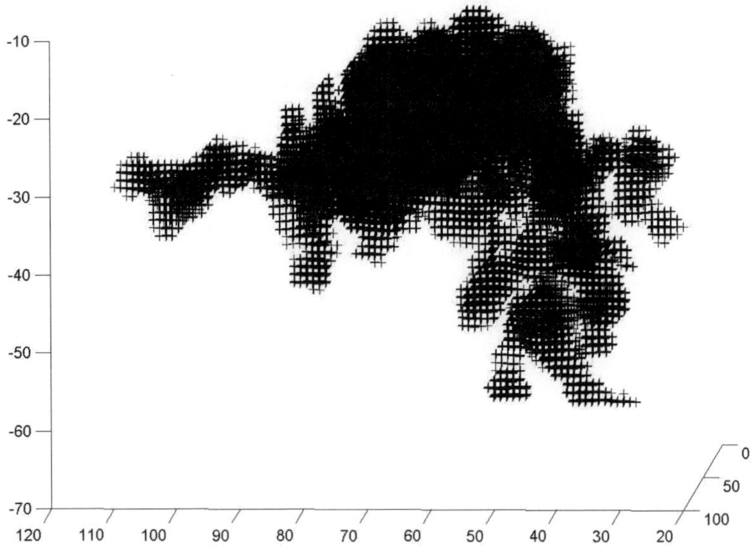

Figure 6.11. Segmented region corresponding to a leaning cylinder object.

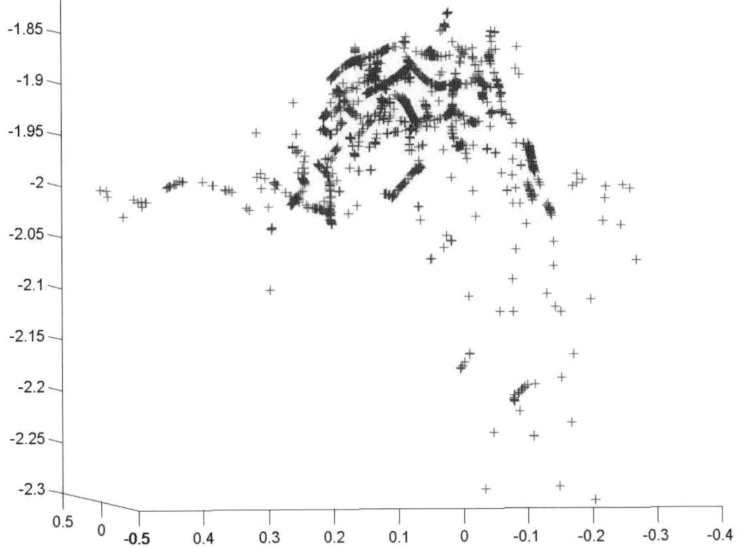

Figure 6.12. Skeleton extracted by the distribution of points of Figure 6.11.

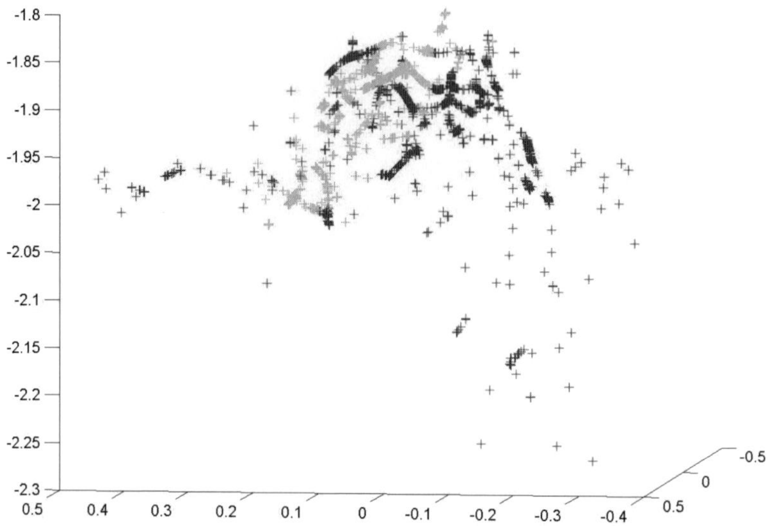

Figure 6.13. The upper base (light grey) of the cylinder has been selected starting with the skeleton of Figure 6.12.

conditions:

- it contains the highest point of the distribution, $P_S = (x_s, y_s, z_s)$;
- it presents a thickness which is related to the system resolution, *ris*, and thus, it is composed of the points that are placed between the planes:

$$\left.\begin{array}{l}\sin(\theta)\cdot\cos(\varphi)\cdot[x-x_S]+\sin(\theta)\cdot\sin(\varphi)\cdot[y-y_S]+\cos(\theta)\cdot[z-(z_S+ris/2)]=0\\ \sin(\theta)\cdot\cos(\varphi)\cdot[x-x_S]+\sin(\theta)\cdot\sin(\varphi)\cdot[y-y_S]+\cos(\theta)\cdot[z-(z_S-ris/2)]=0\end{array}\right\} \tag{6.42}$$

where φ is supposed known, whereas θ is the parameter we want to determine. By assigning different θ we find different planes; the "best" θ is the one that allows us to include the greatest number of points of the distribution between the considered planes. It is then possible to improve the measures of the Euler angles by applying the inertial tensor method to the extracted region. Figure 6.13 shows the same skeleton as Figure 6.12 where the extracted base has been painted light grey.

6.10 CYLINDER MODEL

Once the parameters defining the pose of the analysed object have been computed, the successive step concerns the estimation of features related to the physical dimensions of the target. This task is strongly dependent on the geometrical shape of the considered object; thus, we have developed a surface reconstruction method *ad hoc* for

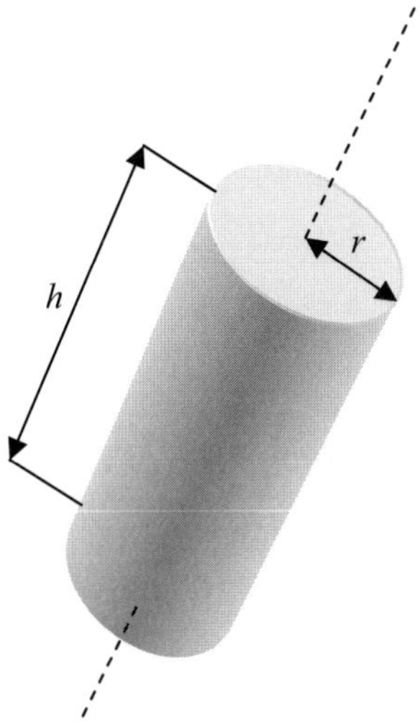

Figure 6.14. Cylinder model parameterized by radius r and height h.

specific object models. We briefly describe the technique for the cylinder model; it is straightforward to extend the procedure to any other shape that can be expressed in an analytical form.

The model refers to a circular right cylinder of height h and radius r. This phase is performed considering the points of the distribution related to the coordinate system placed on the object (OCS), denoted by vectors, $\tilde{\mathbf{v}}_1, \tilde{\mathbf{v}}_2, \tilde{\mathbf{v}}_3$. It is recalled that the OCS has the origin placed in the barycentre of the distribution that generally does not coincide with the true barycentre of the target; in this phase it is thus necessary to update the barycentre estimation by taking advantage of further analyses. In Figure 6.14 the adopted cylinder model is represented.

As in underwater acoustics the scattered energy is totally produced by the object's surfaces, the assumed model is based on the assumption that all the points of the distribution are placed at the same distance (equal to r) from the rotation axis, except for the points located on the eventual upper base of the cylinder that will be contained in a circle of radius r. To obtain an estimation of the radius r, we can project all the points onto a plane normal to the cylinder axis and interpolate the projections with a circumference. Actually, if the projected points are too spread out, it can be useful for an operator to interact by mouse-clicking to select the points belonging to such a circumference. At the end of this procedure we obtain a set of coordinates (x_i^L, y_i^L) corresponding to points placed at an average distance r from the axis of the cylinder. It is now possible to estimate the radius r and the centre coordinates (x_C, y_C) of the circumference that best interpolates the selected points.

The generic equation of a circumference with radius r and centre (x_C, y_C) is the following:

$$(x - x_C)^2 + (y - y_C)^2 = r^2 \tag{6.43}$$

that can be written also in this way:

$$x^2 + y^2 + a \cdot x + b \cdot y + c = 0 \tag{6.44}$$

where the coefficients a, b, c are equal to:

$$\left.\begin{aligned} a &= -2 \cdot x_C \\ b &= -2 \cdot y_C \\ c &= x_C^2 + y_C^2 - r^2 \end{aligned}\right\} \tag{6.45}$$

We can apply Equation (6.44) to the M points denoted by (x_i^L, y_i^L), thus obtaining a linear system of M equations and three unknown variables a, b, c, where $M \gg 3$. In general, such a system is over-determined and there is no solution. In this case it is possible to find a least squares solution, obtaining an estimation for the unknown parameters a, b and c.

Finally, the coordinates of the barycentre projected onto the plane orthogonal to the cylinder axis are equal to:

$$(x_C, y_C) = \left(-\frac{a}{2}, -\frac{b}{2}\right) \tag{6.46}$$

whereas the radius of the cylinder is determined by the following expression:

$$r = \sqrt{\left(-\frac{a}{2}\right)^2 + \left(-\frac{b}{2}\right)^2 - c} \tag{6.47}$$

We have completed the analysis on the plane normal to the cylinder axis; now we must get some information about the axis itself. To estimate the height of the cylinder the simplest method consists in computing the maximum and minimum heights (q_{max} and q_{min}) of the points along the cylinder axis. With respect to the OCS such heights can be expressed by:

$$\left.\begin{aligned} q_{max} &= \max_i \{\tilde{x}_i\} \\ q_{min} &= \max_i \{\tilde{x}_i\} \end{aligned}\right\} \tag{6.48}$$

Therefore, h is equal to:

$$h = q_{max} - q_{min} \tag{6.49}$$

In an analogous way, the height of the barycentre along the symmetry axis can be computed as the average value between q_{min} and q_{max}:

$$q_C = \frac{q_{min} + q_{max}}{2} \tag{6.50}$$

Starting with the coordinates of the barycentre $\mathbf{b}_C = (q_C, x_C, y_C)$ with respect to the OCS – considering the vector inclination $\mathbf{t}$ (6.19) – and the radius and height estimations r and h, we are able to unambiguously represent the cylinder.

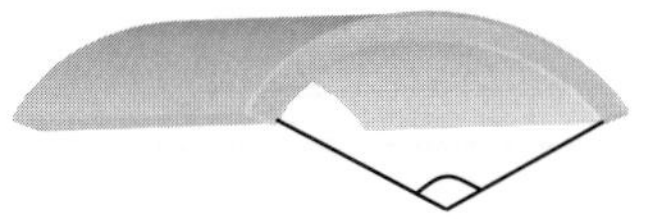

Figure 6.15. Cylinder slice model.

Actually, the developed method provides for creating a model which is also a portion of the cylinder. For example, in the event the segmented region corresponds to a longitudinal slice of a cylinder surface (see Figure 6.15), the user can choose to visualize the whole reconstructed cylinder or only the relative portion.

6.11 VRML RENDERING

The development of an efficient technique of three-dimensional visualization can be a useful step in the human interpretation of both segmentation results and the respective model reconstructions. The operator can explore the ensonified scene and take decisions supported by a better understanding of the objects present in the area of interest. To this aim we have performed a 3D rendering by means of the Virtual Reality Modeling Language, VRML. The VRML format allows one to build a file standardized by ISO/IEC to describe volumetric interactive scenes and worlds. This standard is especially good at visualizing point distributions as segmented data and synthetic representation as reconstructed data, allowing us to explore the obtained results in an accurate way.

By means of VRML a virtual world can be easily described using a classic editing application, by adopting semantics independent of the physical devices or any other implementation-dependent concepts. VRML is also intended to be a universal interchange format for integrated 3D graphics and multimedia. VRML browsers, as well as authoring tools for the creation of VRML files, are widely available for many different platforms. The separation of the descriptive and implementative aspects allows a VRML file containing the information relative to a complex virtual world to be very compact and easily shared on the web. Furthermore, the rendering quality achievable with VRML is very high.

Concerning the visualization of segmented data and object models, VRML offers further advantages. In a VRML file it is possible to define an inertial coordinate system and thus refer all the object local systems with respect to it. This allows one to describe in a simple way the translations and the rotations of the segmented objects. Besides, the VRML semantic already provides the definitions of some geometric primitives – as cylinders, cones, boxes and spheres – allowing one to describe directly some object models. Moreover, it is possible to add new types (called prototypes) which are not provided by the standard semantic and useful to describe more complex shapes, such as truncated cones and rings. Both the standard primitives and the prototypes are parameterized and thus can be adapted to the characteristics of the objects (Figure 6.16).

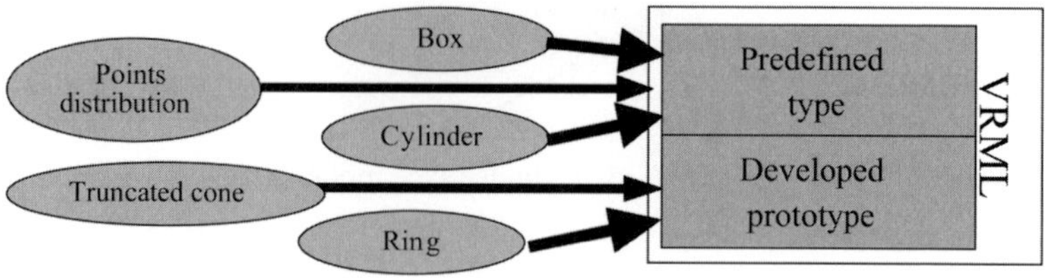

Figure 6.16. VRML standard primitives and prototypes.

6.12 CONCLUSIONS

In this chapter, a complete processing tool of use in the examination of 3D acoustic sub-bottom images has been described. The processing chain allows one to separate – from raw data – the image regions representing natural or artificial objects buried beneath the seafloor. In particular, the devised segmentation process is based on a semi-automatic "seeded" volume growing approach. The voxel classification is guided by a statistical criterion by fitting current volume histograms with an adequate probability density function. The segmented object is then analysed to extract measurements about its shape and pose and to obtain a 3D virtual representation by VRML modellling. In addition, a pre-processing noise reduction stage and a multi-resolution data representation based on an octree approach can be applied, if necessary.

7

Geographic Information Systems for risk assessment of seabed dumping sites

M. Robba, R. Sacile, P. Jonsson and L. Gelaziene

7.1 INTRODUCTION

The preservation of the marine environment affected by dumpsites is strictly connected to the individuation of the sources of pollution and to the assessment of their effects on sea life, human health and economy. To reach this target, a vast amount of information about the status of the environment, together with innovative technological instruments for data collection and techniques for data processing and analysis are needed. This amount of information serves the purpose of helping investigation of the different aspects of the system (toxicity, chemical/physical data, acoustic images, maps, geographic characteristics, sediment status, etc.) that need to be integrated in order to help end-users, on one side, to get a clear view of the environmental problems of the site and, on the other side, to find appropriate solutions from a planning perspective. Specifically, there is a strong need to find instruments that can easily display and organize a great amount of information in order to have an immediate view of the environment status and to facilitate the decisional process.

Geographic information systems (GIS) and decision support systems (DSS) can generate effective and flexible tools to satisfy the above needs. The use of these tools can typically be appropriate whenever it is possible to obtain a quantitative formalization of the problem, adequately supported by engineering and economic evaluations, decision methodologies, and dynamic models of the studied systems. According to the most recent developments in information technology and computer science, the former concept of GIS as a viewer of "static" and "passive" stored information is increasingly evolving towards a new concept of GIS, where additional functionalities allow the computation of a vast quantity of data providing new information and evaluations related to the different aspects of the problem. A functionality that is increasingly requested in a GIS is to provide decision support as related to the problems of concern, such as land and environmental management. These problems are usually characterized by different, multiple and contrasting objectives (such as economic objectives versus environmental protection objectives) that should be harmonized, taking into account heterogeneous and vast databases. In this respect a GIS can be applied to solve a large number of problems related to information management.

The peculiarity of a GIS with respect to other information systems is that data are geo-referenced and are, in general, heterogeneous, as referred to different technical, chemical, physical or biological aspects. In addition, the data acquisition processes are, by definition, distributed. As a consequence, professionals and researchers working on GIS applied to environmental management should have expertise in several scientific methodologies applied to the territory, such as those concerning chemical, physical or biological aspects as well as computer science and decision-related methodologies.

A GIS can be considered a layered information system. Each layer can offer a reliable service to higher layers, and the link between adjacent layers is guaranteed by an adequate communication service (this statement implies that, at least in principle, adjacent links of the same GIS layer can be spatially distributed).

The first level, from bottom to top, can be defined as the "environmental monitoring layer", whose task is to acquire information from the environment by the use of sensors; in a broad sense, included in this definition are, for example, rain gauges, anemometers, cameras for map data acquisition from satellites, water sample collecting teams and related water quality analysis processes. In the "environmental monitoring" layer, sensors and related software for data acquisition and communication transfer are present and distributed in space, providing quite different content and formats of heterogeneous data.

The second level can be defined as the "Data Base Management System (DBMS) layer". In this layer, data coming from telematic connections or from storage supports are modelled and stored in one or more databases. The most frequently used data models are based on the relational paradigm, and these data are often centralized – for example, in local or regional authorities.

The third level can be defined as a "data analysis, synthesis and processing layer", and it should produce "simulations" either to predict future behaviour of the monitored environmental system, or structural changes, when new anthropic or environmental scenarios are imagined. A GIS often takes the role of decision support system (DSS), in the sense that, with its ability to condense a large amount of data in a single map, it can provide reliable and effective support to a decision maker. However, a fourth layer – that is not commonly present in a true GIS – will be inceasingly requested for a modern GIS: one that offers additional capabilities such as mathematical programming computation.

In the literature, the application of computers to information about the terrestrial surface is widely used and well known (Wright and Bartlett, 1999), while for the marine and coastal environment things are much less straightforward. From the perspective of geographic information science, the growing interest in marine and coastal applications is enormously fascinating. The increased interest in oceans has brought about improvements in the analytical potential of GIS, while extending the methodology framework for marine applications (Breman, 2002). Many challenges remain, such as addressing the multiple dimensionality and dynamism of oceanographic data and the development of effective conceptual and data models of marine objects and phenomena. In Breman (2002) is described how GIS capabilities have been used to represent different aspects connected to the marine environment.

Within the SITAR project, GIS development has been mainly directed toward the goal of supporting decision makers to access the great amount and diversity of information involved in the risk assessment of dumpsites. The general design guideline has been that of allowing the user to start from general thematic maps, where the information is presented graphically in a qualitative manner that is intuitively appealing. At user demand, the system can display information with an increasing level of detail, ending up with raw data points, should the user wish to inspect them. Economical, social and political considerations – in particular, related to the management of the site – are out of the scope of the project, and such information has not been included in the developed system.

7.2 SYSTEM ARCHITECTURE

The system specifications have been developed with the active involvement of environmental end-users in the design phase; throughout the evolution of the project, partial implementation of the system has been demonstrated to end-users, in order to receive their feedback, and, whenever possible, the original design has been modified according to the new indications. The resulting system architecture (represented in Figure 7.1) is based on the Data Base Management System (DBMS), developed in MS Access 2000, and the geographic information systems (GIS), developed in ArcGis 8.1

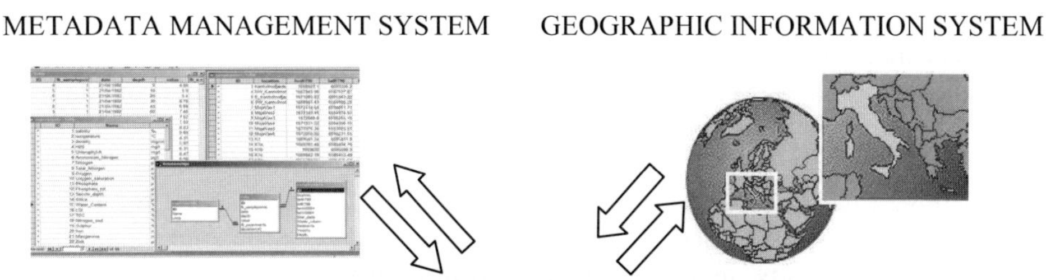

Figure 7.1. The system architecture: a Data Base Management System and geographical information are merged in a GIS interface.

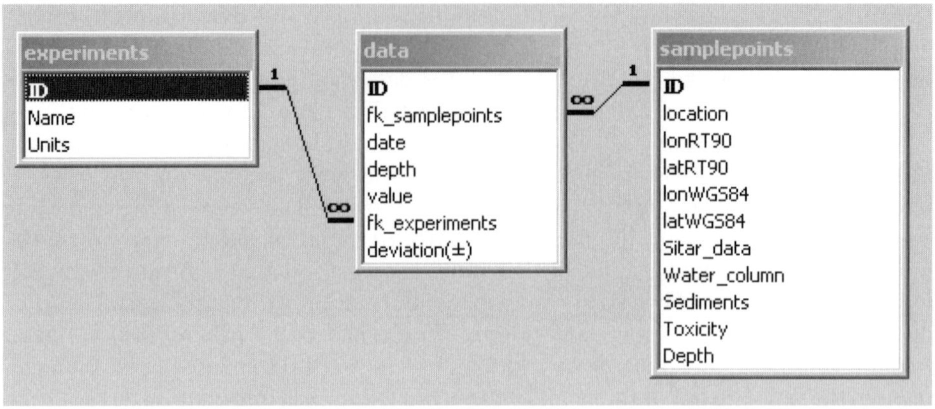

Figure 7.2. Relations among tables in the developed database.

software layers, integrated by means of a graphic interface, developed in Visual Basic 6.0 using ESRI MapObjects 2.0 libraries. Several data classes can be handled (metadata, maps, videos, images, etc.) and different kinds of queries can be performed according to end-user requirements. The data that have been collected and that are necessary to describe the status of the dumpsite can be divided into four main sets:

(1) environmental data: that is, oceanographic information and chemical–physical parameters of the water column and the sediments;
(2) toxicological data: that is, data from nanoinjection technique tests;
(3) sea bed data: that is, data from standard bathymetry, side-scan sonar, sub-bottom profiling surveys as well as data from SITAR-investigated parametric side-scan sonar equipment;
(4) object image data: that is, standard video survey of barrels exposed over the seabed, as well as SITAR-generated 3D reconstructions of buried containers (from multiple-aspect survey data).

The relational database, realized in MS Access 2000, is characterized by different tables that relate the numerical value, the type of experiments that have been performed (i.e., salinity, temperature, embryo mortality, water content, etc.), their units, longitude and latitude (in RT90 and WGS84), depth of the bottom from the surface, the type of data (SITAR, water column, sediment, toxicity), and object description. Figure 7.2 reports an example of relations among the different tables.

The database is composed of five tables: *data*, *experiment*, *objects*, *samplepoints* and *subbottomprofile*. The table *data*, *experiment* and *samplepoints* are related and the specific fields are reported in Figure 7.2. In table *data* all data are stored with characteristics expressed by the following fields: foreign key[1] to the sample points,

[1] A foreign key is a field in a database record that points to a key field, forming a key of another database record in some different table.

date, depth, value, deviation and foreign key to the type of experiment. The table *experiment* relates to the parameters that have been measured during the experimental campaigns. The characteristics of the various experiments performed is described through two fields: the name of the experiment (salinity, temperature, etc.) and the units in which the data have been measured. Table *samplepoints* report the description of the points on which the experiments were performed. The fields are: location (name of the station), lonRT90 (longitude in Swedish RT90), latRT90 (latitude in Swedish RT90), lonWGS84, latWGS84, sitar_data (yes/no), water_column (yes/no), sediments (yes/no), toxicity (yes/no), depth of the bottom. Table *objects* report the description of the objects that have been found in the study area and that might be dangerous for marine ecosystem, with regard to the files related to images, videos, VRML models, etc. Finally, in the table *subbottomprofile*, the fields report the coordinates of the points that compose the line traced by the ships during the experimental campaigns.

The information in the database can be retrieved by specific queries (selected according to end-user specifications), the results of which are displayed by means of charts. The queries are accessible through the developed interface, allowing information about the water column data, sediments, the toxicity analysis and objects on the seabed to be obtained. The location of each sample point is displayed in a specific map layer. The data values, as well as the images and videos relative to a specific point or line, are directly accessible by simply clicking on the map layers. An additional pop-up mask allows the user to access, should it be needed, structured information – such as evolution in time, correlation with other data points, etc. (Figure 7.3).

The layers created in the GIS represent different maps that are viewed through the interface (seen in different projections and coordinate systems: WGS84, RT90). The user can simultaneously or separately see the following layers by clicking directly on the interface: coastline, ocean, bathymetry, side-scan sonar, sub-bottom profiles, toxicity sample points, sediment sample points, water column sample points and object sample points. Figure 7.4 (see colour section) shows how the object layer, the side-scan sonar layer and the sub-bottom profile layer are displayed together on the interface. As an example, interrogating (by clicking) the sub-bottom profiles lines reveals the corresponding data.

The experimental campaigns performed for the SITAR project gave the possibility to analyse different samples in the study area. Specifically, several physical, chemical and toxicity parameters can be associated to different sample points in the area. SITAR thematic maps geographically compare the values obtained in the experimental campaigns and in the laboratory analysis. In particular, for the toxicological data colour-coding has been employed, with hot colours (e.g., red) corresponding to the level associated to higher relative toxicity and cold colours (e.g., blue) corresponding to lower relative toxicity. A legend reporting the quantitative values of the map is always displayed, giving the user the possibility of defining the dynamic range of the scale displayed. Thematic maps can thus be constructed for various toxicity biomarkers; in Figure 7.5 (see colour section) an example is shown from the SITAR experimental area – the displayed data are reported for illustrative purposes only while their environmental significance is discussed in Chapter 15.

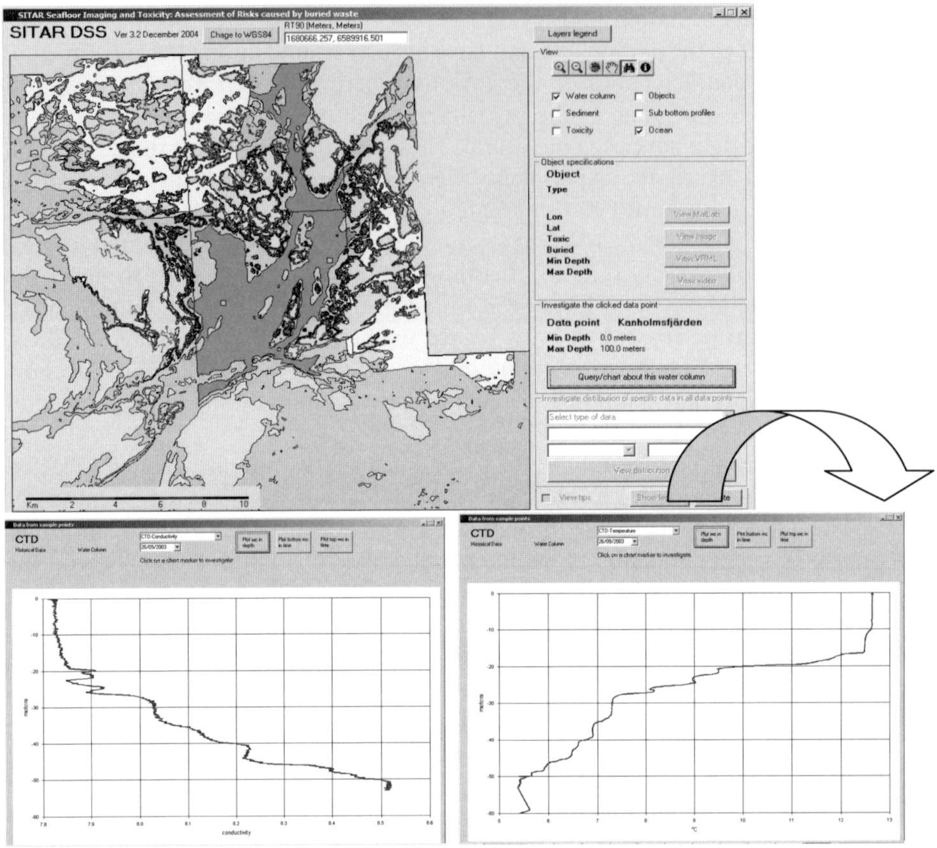

Figure 7.3. An example of a query about conductivity and temperature: such as charts at a certain depth at a specific date and sample points on data points (water column data, salinity versus depth at selected date). Interrogation is made by clicking on the data point on the map. A pop-up window allows a number of different queries on the data point to be selected.

7.3 CONCLUSIONS

The use of geographic information systems in marine environmental applications can be very helpful in order to display data in an understandable way and in order to support the decision making process. The SITAR DSS is heavily information-oriented. The main goal of its implementation is to support decision makers to access a great amount of heterogeneous information concerning marine dumpsites, analysing and synthesizing it in a GIS-oriented interface. Specifically, metadata are organized in a relational database (MS Access 2000), while geo-referenced information has been organized using a GIS tool (ESRI Arcview 8.1). All these data are accessible through a graphical user interface (GUI).

Part III

Experimental activities

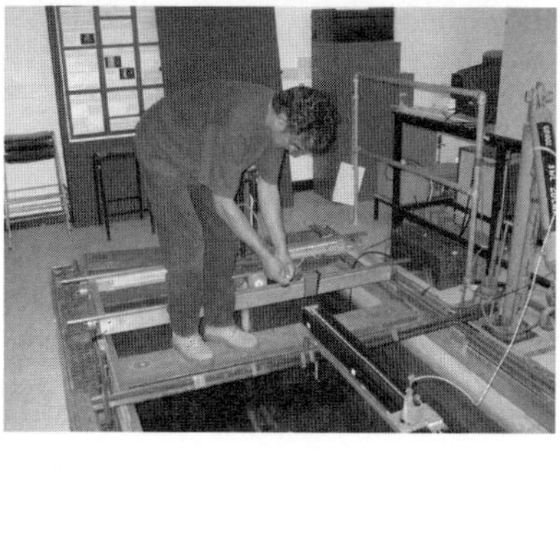

8

Scaled multiple-aspect scattering tank experiments

Ph. Blondel, N. Jayasundere, P. F. Dobbins and M. Cosci

8.1 SCALED EXPERIMENTS – SET-UP

The first objective of the scaled experiments was to provide reliable acoustic scattering data in a controlled environment, in order to test the techniques and predictions of the numerical models developed as part of the SITAR project (see Chapter 3). A second objective was to form a data base for analysis and investigation of acoustic scattering from single and multiple targets, proud and partly/fully buried in different types of seabed, which would be of use to researchers within the SITAR project and worldwide.

The experimental set-up was designed to be a scaled-down version of the one to be used during the SITAR sea trials, which took place in September/October 2003 in the Stockholm Archipelago (see Chapter 11). Systematic and thorough measurements were carried out as a function of the incidence angle, the scattering angle and the bistatic angle (see Figure 8.2). A portion of the results is presented here, showing the influence of the number of targets and their respective positions. These experiments used the University of Bath underground tank facilities (Figure 8.1). These were set up in the 1970s and provide a controlled environment with different types of real seabeds, carefully prepared and emplaced at the bottom of the tank (N. G. Pace, pers. commun., 1999).

Owing to both the technical constraints of the Bath facilities (tank dimensions, seabed types and transducers available) and the experimental constraints when preparing the sea trials (seabed types and transducers available), it was decided to use a scaling factor of approximately 10:1 in the design of the experiments.

The targets were designed to match the numerical models used during SITAR (Skogqvist and Karasalo, 2003) and the types of targets likely to be encountered at the trial site. They are listed in Table 8.1. Target T_1, for example, is an approximation of a scaled oil drum, whose standard dimensions are 58-cm diameter and 88.5-cm height. Target T_5 is used for comparison with the numerical simulations of Skogqvist and

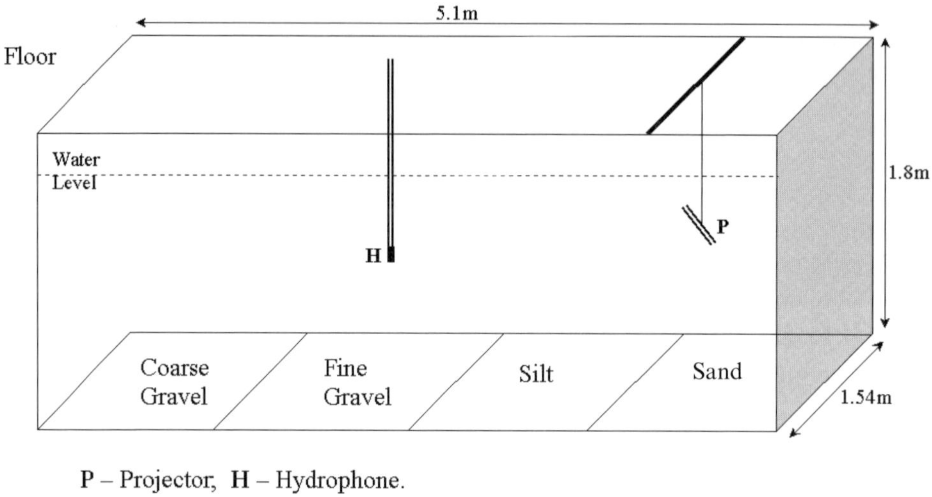

Figure 8.1. The University of Bath tank facilities. The walls are made of concrete, and the top of the tank is at floor level. The sediments in the trays are 14 cm deep on average. The water level can be varied. Both acoustic projector and hydrophone(s) can be positioned anywhere in the tank. The targets are oriented relative to the x-axis (lengthwise), the y-axis (along the 1.54-m width of the tank) or the z-axis (vertically). They were also placed diagonally (x–y).

Karasalo (2003). The SITAR sea trials also showed the presence of a large, metal box (Figure 8.2). A scaled-down version (target T_6) was therefore manufactured and used. All six targets were placed proud, half-buried or flush buried, at different orientations, either alone or in groups.

The four sediment trays at the bottom of the tank were uniformly filled with sediments typical of European continental margins. The depth of sediment varied slightly from tray to tray. Those used in this study were filled with silt (50-μm mean grain size) and gravel (5-mm mean grain size). The former are a scaled-down version of the soft muddy sediments, with a minute content of gas, expected at the sea trial site

Table 8.1. Dimensions of the scaled targets used in the tank experiments: diameters (D), lengths (L) and other characteristics influencing their acoustic scattering properties.

T_1	Sealed aluminium tin, fluid-filled, 6.7 cm (D) × 10 cm (L)
T_2	Air-filled stainless steel cylinder, 5.8 cm (D) × 10.4 cm (L), sidewall thickness 3 mm, end cap thickness 2 mm
T_3	Solid aluminium cylinder, 5.1 cm (D) × 8.1 cm (L)
T_4	Solid steel cylinder, 7 cm (D) × 8 cm (L)
T_5	Solid brass ring, 10.5 cm (D) × 7.5 cm (L), wall thickness 7.5 mm
T_6	Air-filled stainless steel box, 10 cm × 5 cm × 5 cm, wall thickness 2–3 mm

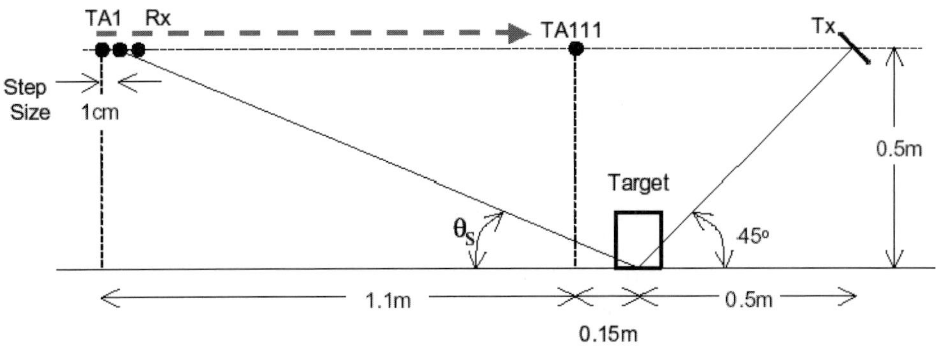

Figure 8.2. A typical line scan. The hydrophone Rx moves in 1-cm steps from position TA1 to TA111, giving a range of scattering angles θ_s. The transmitter Tx is tilted at 45° in this case.

(as confirmed later). Karasalo and Skoqvist (2004) used results from previous inversions at nearby sites to obtain tentative values of 1.047 for the sound speed ratio and 1.1 for the density of the sediments, quite close to the values for the silt tray (1.024 and 1.204, respectively). Only the results for the silt tray are presented here. The sediment depth inside the silt tray is ca. 14 cm, giving a theoretical attenuation of 184 dB/m at 238 kHz.

8.2 DATA ACQUISITION AND PROCESSING

The experiment was arranged according to the set-up of Figure 8.1. The height of the water column was 1.475 m. The acoustic projector was a damped piezoelectric transducer resonant at 238 kHz, excited with a single sinusoid at 238 kHz with a 20-V peak-to-peak amplitude (although other signals – e.g., chirps[1] – were also used later). It had a narrow beamwidth (about 10°), measured in 2D and in 3D (Blondel *et al.*, 2001; Jayasundere and Blondel, 2004). The 3D calibration shows non-symmetrical variations for the angles further away from the axis of the transducer; this needs to be remembered when analysing scattering from the larger angles. The acoustic projector was mounted on a gantry supported across the tank, allowing its tilt to be set to any desired angle. The transducer was positioned 0.5 m away from the target(s). The receiver is an omnidirectional hydrophone Brüel & Kjaer (B&K) 8103, fixed to the end of a 9-mm diameter stainless steel tube. The tube is mounted on a translation/rotation stage, positioned in line with the projector and the target. Two types of experiments were performed: line scans and rotation scans. In line scans (Figure 8.2), the hydrophone was positioned 0.5 m above the seabed, and remotely translated via a PC with a step size of 1 cm, usually starting 1.1 m away from the target, and finishing 15 cm from the target. In rotation scans, the tube containing the hydrophone was at a fixed distance of 0.5 m from the main target and remotely rotated via a PC, sampling

[1] A chirp is a time-limited sinusoidal signal.

the scattered acoustic field as a function of bistatic angles between 140° and 220° (180° corresponding to in-plane scattering), with a 2.5° step size. The field was sampled as a function of the scattering angle by manually adjusting the height of the hydrophone over the seabed from 1.375 m to 0.15 m with a step size of 2.5 cm.

These two types of scans correspond to the typical investigation strategies adopted when using ROVs at sea, either "flying lines" over a region or circling objects of interest. They also offer a broader sample of possible bistatic and scattering angles, despite the limitations brought by the finite size of the tank. The total range of configurations sampled is presented in Table 8.2. The acoustic field was also measured for the bare seabed, before placing the targets and after removing them (to assess any disturbance).

Typical results are presented in Figure 8.3; they show the variations of acoustic signals as a function of time, and the significant influence of the target as the scattering angle varies. Other forms of data visualization have been explored, looking at variations as a function of the bistatic angle and the incidence angle. Frequency-domain analyses (e.g., short-term Fourier transforms) have also been performed (Blondel *et al.*, 2006).

8.3 TARGET IDENTIFICATION AND RECONSTRUCTION

Multiple targets can be visually identified by looking at the waveforms in the time-domain at different scattering angles, and careful use of different bistatic angles is often desirable. Both the tank experiments and, later, the sea trials, used a variety of transmitted signals (Ricker[2] pulses, chirps, etc.). Reliable techniques for the systematic identification of targets need to be coupled with exact derivation of their positions/sizes, to enable target reconstruction. Existing target identification techniques work well with one kind of signal (e.g., chirp) but generally not with another (e.g., Ricker pulse); they are also very sensitive to noise. Advanced deconvolution methods and search routines were therefore developed to reconstruct acoustic scattering inside and outside the objects. Dobbins *et al.* (2003) showed these scatterers could be detected even in high-noise signals, using an adaptive implementation of the Wiener filters used in seismic data processing. Once processed, these signals can be used for localization of the target(s), using multidimensional search routines based on a downhill simplex algorithm (Dobbins and Blondel, 2004). Figure 8.4 (see colour section) presents typical results from the tank experiments, showing the very good agreement between the inferred and actual locations of the targets.

[2] A Ricker signal is a specific, pulse-like, signal shape very often used in marine seismic exploration.

Table 8.2. Range of bistatic scattering configurations sampled during this study (silt tray only; a similar dataset was acquired using the gravel tray).

Target	Configuration	Bistatic angles (°)	Axis orientation
Line scans: incidence angle 45°, scattering angles 15°–75° (variable according to configuration)			
Bare seabed	N/A	160, 170, 180, 190, 200	
T_1	Proud	160, 170, 180, 190, 200	x, xy, y and z
	Half-buried	160, 170, 180, 190, 200	x and z
	Flush-buried	160, 170, 180, 190, 200	z
T_2	Proud	160, 170, 180, 190, 200	x, xy, y and z
	Half-buried	160, 170, 180, 190, 200	x and z
	Flush-buried	160, 170, 180, 190, 200	x and z
T_3	Proud	160, 170, 180, 190, 200	x, xy, y and z
	Half-buried	160, 170, 180, 190, 200	z
	Flush-buried	160, 170, 180, 190, 200	z
T_4	Proud	160, 170, 180, 190, 200	x, xy, y and z
T_5	Proud	160, 170, 180, 190, 200	z
	Half-buried	160, 170, 180, 190, 200	z
	Flush-buried	160, 170, 180, 190, 200	z
Line scans: incidence angle 30°, scattering angles 15°–75° (variable according to configuration)			
Bare seabed	N/A	180	
T_1	Proud	180	x and z
T_2	Proud	180	x and z
T_3	Proud	180	z
Rotation scans: incidence angle 45°, scattering angles 14°–70°			
T_1	Proud	140–220 (2.5)	x and y
T_2	Proud	140–220 (2.5)	x
T_2	Flush-buried	140–220 (2.5)	x
T_1+T_2	Proud	140–220 (2.5)	y
$T_1+T_2+T_3$	Proud	140–220 (2.5)	y
$T_1+T_2+T_3+T_4$	Proud	140–220 (2.5)	y

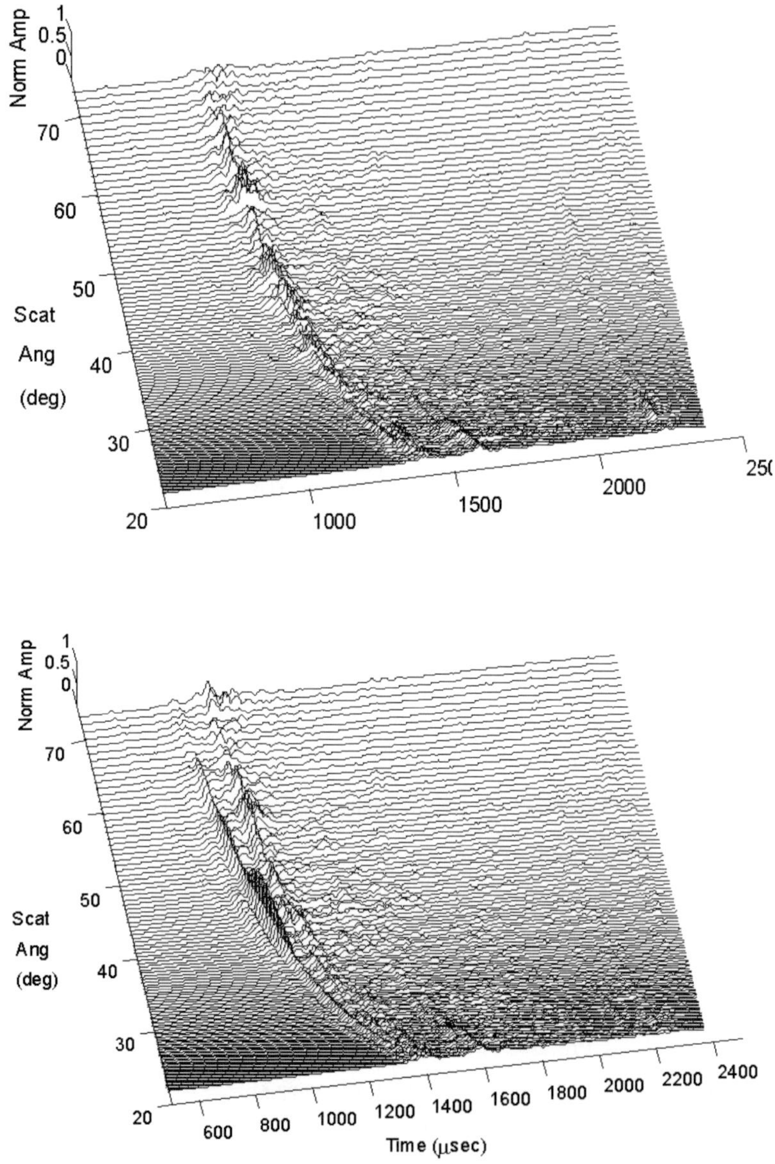

Figure 8.3. Scattered waveforms for bare silt (top) and with target T_1 proud in the z-direction – that is, placed vertically on the seabed (bottom). In this case the incidence angle is 45° and the bistatic angle is 180° (in-plane scattering). The normalized amplitudes are plotted as a function of time and scattering angle.

9

Full-scale tank parametric side-scan sonar test

M. Zakharia

9.1 SYSTEM DESCRIPTION

As mentioned briefly in Part II, p. 14, the side-scan parametric sonar, built within the framework of the SITAR project, possesses the following specifications:

- The primary frequency is centred on 100 kHz and the secondary frequency ranges from a few kHz up to 20 kHz.
- In order to complete the sonar data by a map of the area, a set of two receiving lines was used for correlation bathymetry processing after SAS. The sonar geometry is described in Figure 9.1.

The sonar is composed of three transmitting channels. Transmitter length (and thus SAS resolution) could be controlled by using one, two or three channels. Two lines of eight receiving channels were used. For every sub-array, two channels (HF and LF) were constructed. In total, for every ping, 32 channels were digitized simultaneously and stored onboard the ship for further processing.

A controlled tank experiment was performed as part of SITAR activities for calibrating the sonar and conducting a mock-up experiment for validating, at scale 1:1, all the algorithms on benchmark data (in addition to validation on simulated data). The experiment was conducted in spring 2004 – that is, *after* the field activities with the same instrument as described in Chapter 11. The tank facilities were offered by GESMA, French Navy, in Brest: they consisted in a full-scale tank which was previously used as a submarine hangar. The equipment was placed in the tank when dry, and then the tank was gradually filled with sea water.

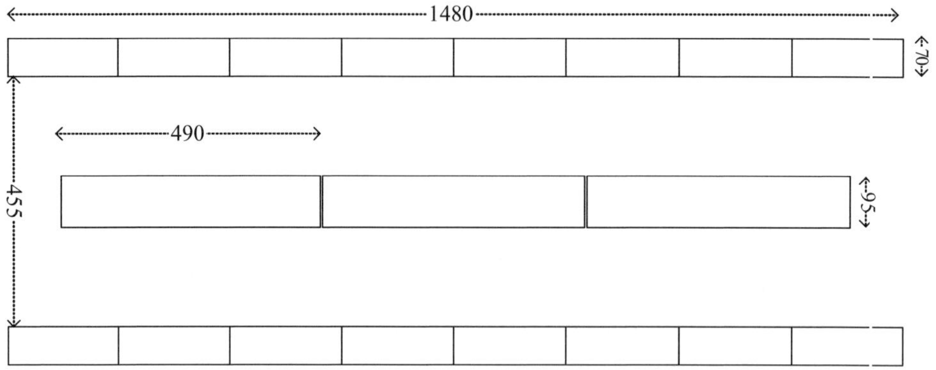

Figure 9.1. Array geometry: (top) upper receiving array (2 × 8 channels); (bottom) lower receiving array (2 × 8 channels); and (centre) transmitting array (three channels).

9.2 SYSTEM CALIBRATION

Calibration of the transmitter and, in particular, the measurement of its directivity pattern pointed out a hardware problem. This problem was due to an error in wiring on the connectors and was repaired by KDA before conducting the tank experiments. After the wiring problem was solved, the new directivity pattern was measured (Figure 9.2).

The source levels obtained after the repair was 218 dB re μPa @ 1m for the primary frequency and 175 dB re μPa @ 1 m for a secondary frequency of 20 kHz. The source level for the secondary signal was lower (172 dB) at lower frequencies (10 kHz).

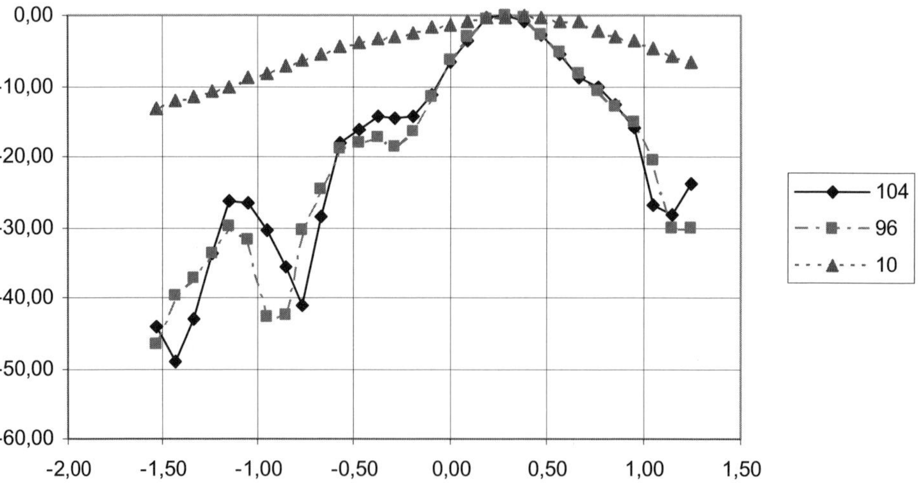

Figure 9.2. Directivity pattern of parametric transmitter after repair: x-axis in degrees, y-axis is normalized amplitude (in dB).

9.3 SAND BOX EXPERIMENT

The experiment was run using a sand box containing several targets as shown in Figures 9.3 and 9.4.

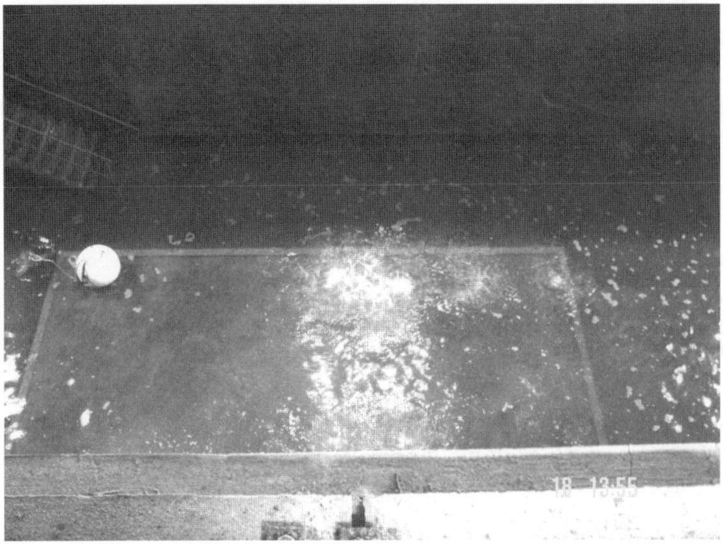

Figure 9.3. Sand box photograph.

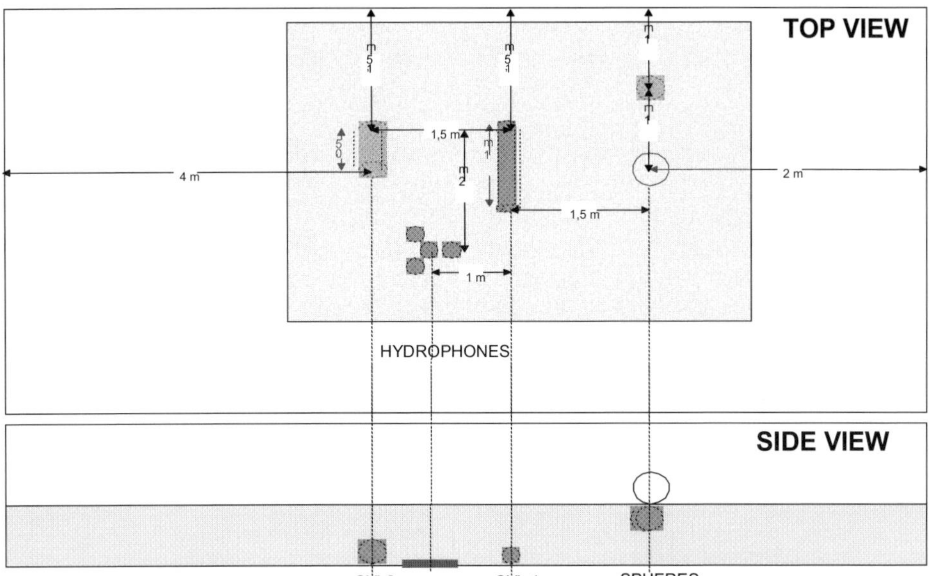

Figure 9.4. Layout of the test targets in the sand box.

Time series data were recorded as a function of position across the test tank. After the experiment was finished, the data were processed. The targets comprised:

- Two hollow cylinders (CYL1 and CYL2), one laid on 5 cm of sand and another buried in the sand.
- Solid sphere no. 1 (∅ 10 cm), flush buried
- Solid sphere no. 2 (∅ 15 cm) on top of the sand surface.

Hydrophones were also buried in order to measure transmission in the sand. Several side-scan geometries were used and data were acquired either continuously or position by position.

9.4 RESULTS

All the following figures concerning the tank data possess the same scales and the same dynamic range as the ones shown in Figure 9.4. In this figure the positions of the targets are also shown.

Figures 9.5 to 9.8 (all in colour section) show the interest of SAS focusing in all three bands. The effect is more visible in high frequency as, at this range, the parametric effect is predominant with respect to the synthetic one. The images show that, even under controlled conditions, a correction algorithm such as DPC is required in the high-frequency range.

10

The sea trial site in Möja Söderfjärd: biological sampling

P. Jonsson and T. Hansson

10.1 BACKGROUND

The full-scale sea trial organized within SITAR had the intention of testing different methods in a realistic situation and, at the same time, under the best possible control obtainable with an operation at sea. For this reason, as the SITAR study area a bay in the Stockholm Archipelago, Möja Söderfjärd, 45 km E of Stockholm, was selected. This site witnessed the dumping of ammunition and civilian waste in the period 1900–1965. The site is environmentally well documented (from surveys of SEPA, Swedish Navy, Swedish Geological Institute), showing the presence of steel and wooden barrels resting buried or half-buried in the sediment. Documented dumping shows the presence of objects of comparable sizes with those to be investigated in the project. Water depth in the area varies between 75 to 105 metres. A wealth of historical data collected by the above-mentioned institutions was made available to the project for inclusion in the GIS-integrated data presentation system, and in order to "ground-truth" the SITAR findings.

10.2 BIOLOGICAL SAMPLING FOR NANOINJECTION TECHNIQUES

After identification of the study area, the reference site for biological sampling was chosen at Möja Västerfjärd, also in the Stockholm Archipelago. Möja Västerfjärd is 2–5 nautical miles north of Möja Söderfjärd, its bottom is affected by the same geological and sedimentary processes that characterize the experimental site at Möja Söderfjärd and, since the prevailing currents in the area flow from north to south, it is not affected by transport of toxic material from the experimental site. For these reasons, Möja Västerfjärd was considered the ideal site for reference purposes. A control site just adjacent to the dumping site was also identified. After identification of the sampling sites, it was decided to collect samples from all the three sites within the

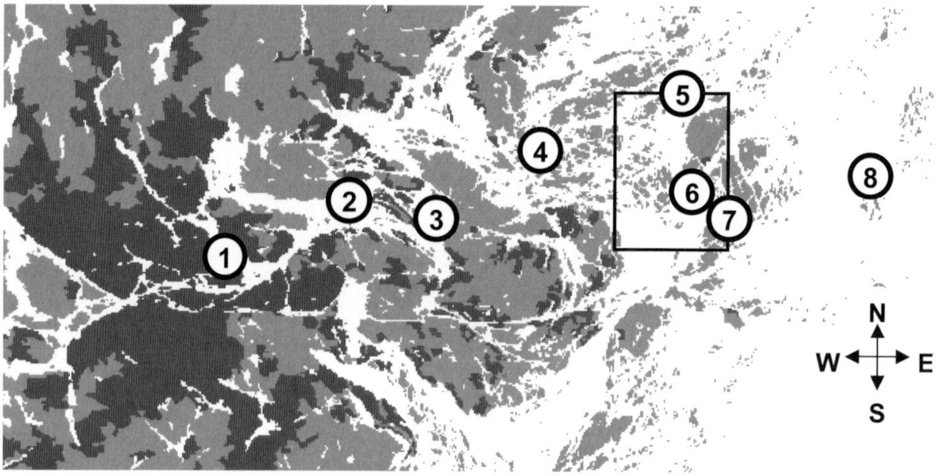

Figure 10.1. The Stockholm Archipelago gradient (Sweden, Baltic Sea) including the SITAR study area (6–7): (1) Waldemarsudde, (2) Tegelön, (3) Torsbyfjärden, (4) Gällnöport, (5) Möja Västerfjärd, (6) Möja Söderfjärd, (7) Lökholmen, (8) Björkskär. White: water. Grey: land. Dark grey: developed area (Stockholm). Inset see Figure 10.2. Scale: 1:550,000.

same operation. Sediment sampling was performed by SU, with the assistance of SEPA, onboard the R/V *Sunbeam* on 11 September 2002. The 21 sampling stations in the Stockholm Archipelago are shown in Figures 10.1 and 10.2. Möja Västerfjärd and Möja Söderfjärd are referred to as the SITAR study area, which is shown in Figure 10.2. Sediment cores were collected at five sites within the munitions dumping area (Möja Söderfjärd 3–7) and along a gradient with two sites south of the dumping area (Möja Söderfjärd 8 and 9) and three sites north of the dumping area (Möja Söderfjärd 1–2 and Möja Västerfjärd 6). Möja Västerfjärd 1–5 sites further north of the dumping area were considered as reference sites. The exact position of all investigated sites is given in Table 10.1. At each site four sediment cores were taken using a Gemini twin corer. The top 5 cm, corresponding to the last 2–3 years of sediment accumulation, were sliced off and the samples from each site were pooled. The samples for the biotoxicity test and chemical analysis were immediately put in a freezer onboard the ship and stored frozen until further preparation. The samples for analysis of water content and total organic carbon (TOC) were put in a refrigerator and stored cold until further preparation.

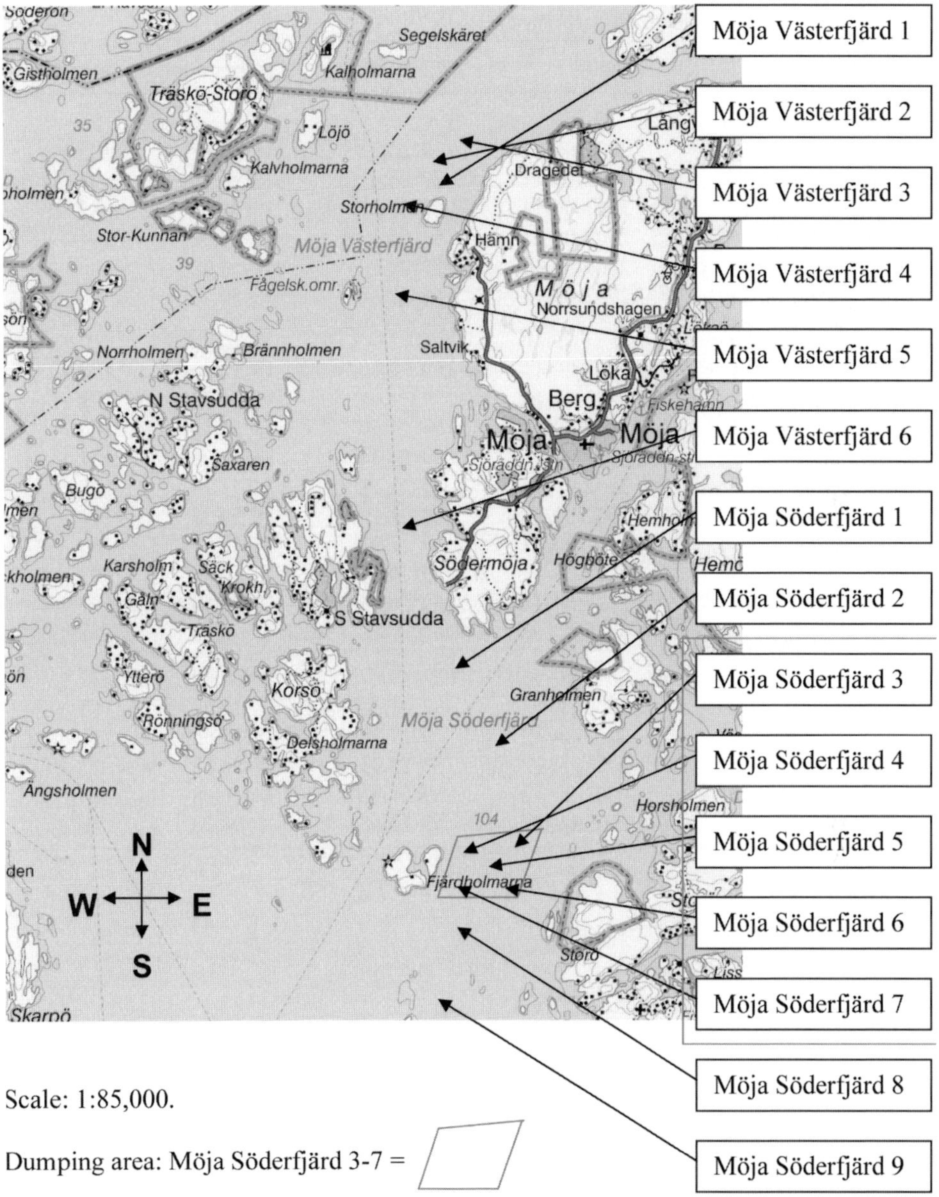

Scale: 1:85,000.

Dumping area: Möja Söderfjärd 3-7 = ▱

Gradient sites: Möja Västerfjärd 6 and Möja Söderfjärd 1, 2, 8 and 9
Reference sites: Möja Västerfjärd 1-5

Figure 10.2. The SITAR study area.

Table 10.1. Positions of the investigated sites.

Station	Date	Latitude	Longitude	Depth [m]
Bay Möja Västerfjärd				
Möja Västerfjärd 1	2002-09-11	59°26.25′	18°50.87′	103
Möja Västerfjärd 2	2002-09-11	59°26.43′	18°50.60′	98
Möja Västerfjärd 3	2002-09-11	59°26.57′	18°51.14′	82
Möja Västerfjärd 4	2002-09-11	59°26.10′	18°50.32′	105
Möja Västerfjärd 5	2002-09-11	59°25.38′	18°50.32′	108
Möja Västerfjärd 6	2002-09-11	59°23.88′	18°50.27′	98
Bay Möja Söderfjärd				
Möja Söderfjärd 1	2002-09-11	59°22.70′	18°50.90′	104
Möja Söderfjärd 2	2002-09-11	59°22.04′	18°51.35′	109
Möja Söderfjärd 3	2002-09-11	59°21.59′	18°51.75′	106
Möja Söderfjärd 4	2002-09-11	59°21.44′	18°51.25′	107
Möja Söderfjärd 5	2002-09-11	59°21.38′	18°51.24′	107
Möja Söderfjärd 6	2002-09-11	59°21.19′	18°51.52′	101
Möja Söderfjärd 7	2002-09-11	59°21.23′	18°50.97′	106
Möja Söderfjärd 8	2002-09-11	59°20.84′	18°50.56′	103
Möja Söderfjärd 9	2002-09-11	59°20.35′	18°50.50′	77
Stockholm Archipelago gradient				
Waldemarsudde	2002-09-13	59°19.07′	18°06.62′	31
Tegelön	2002-09-13	59°21.50′	18°19.50′	15
Torsbyfjärden	2002-09-12	59°19.37′	18°27.66′	31
Gällnöport	2002-09-12	59°23.73′	18°36.65′	27
Lökholmen	2002-09-12	59°20.91′	18°56.68′	54
Björkskär	2002-09-12	59°23.68′	19°08.78′	78

11

Acoustic sea trial in the Möja Söderfjärd dumpsite

P. Moren, A. Caiti, M. Zakharia, M. A. Larsen, Ph. Blondel and J. Dybedal

11.1 TRIAL PLANNING AND ORGANIZATION

The dumpsite in Möja Söderfjärd Bay indicated in Chapter 10 was the location for acoustic tests of the SITAR-developed techniques. Traditional surveys covering the selected dumpsite in the Möja Söderfjärd Bay were, together with other available background information, included in the overview planning of the sea trial. The water depth is around 100 m in the major part of the site, except for the eastern region where the depth decreases to around 70 m. The Swedish Geological Survey (SGU) mapped the area for the SITAR field trials as part of their regional survey campaign. Mapping was accomplished using a 3.5-kHz sub-bottom echo sounder, but – due to high gas content in the muddy sediment – almost no acoustic penetration was achieved. A complementary sub-bottom survey, performed using a towed 3–8-kHz chirp[1] sonar, showed somewhat better sediment penetration, especially in the shallower parts of the site. Still, the resolution of the two traditional sub-bottom sonar systems used was not high enough to resolve small objects on, or buried in, the seabed (Figure 11.1). The bottom of the dumpsite was also mapped using a traditional side-scan sonar (Klein system, 100 kHz). From the mosaic picture (Figure 11.2), SGU reported ca. 450 objects were visible on the seafloor. Many of the localized objects were also identified using a ROV-mounted camera. The exact number of objects completely buried in the soft sediments was unknown.

The SITAR sea experimental activities took place between 22 September and 10 October in Möja Söderfjärd. The trial included two ships, the R/V *Altair* and HMS *Fårösund* (Figure 11.3). The R/V *Altair* operated the parametric side-scan sonar prototype, while HMS *Fårösund* operated two ROVs and the vertical hydrophone array for video inspection and multiple-aspect scattering measurements. Personnel

[1] A chirp is a time-limited sinusoidal signal.

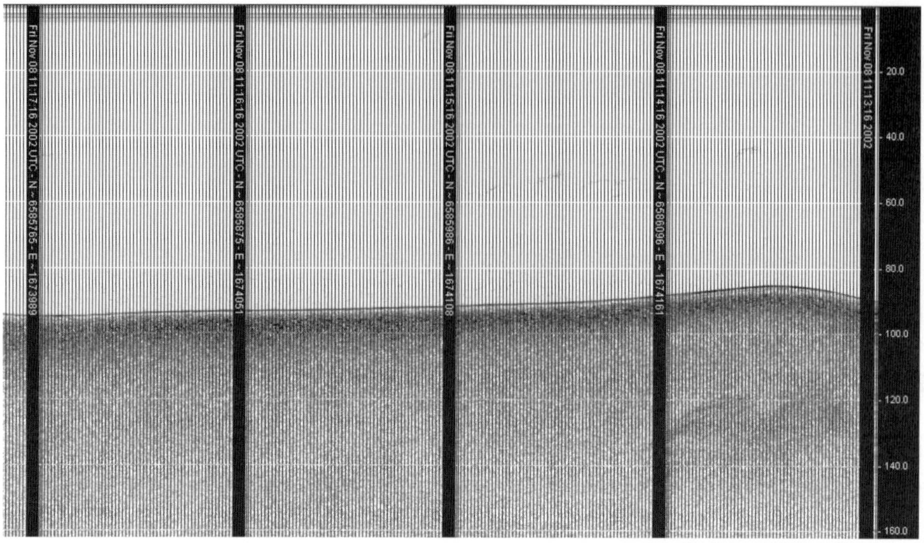

Figure 11.1. Example of a sub-bottom profile from Möja Söderfjärd. The section is 400 m long and was measured using a traditional, towed 3–8-kHz chirp sonar. The poor penetration was due to gas in the sediment and the resolution was not sufficient to resolve any small objects.

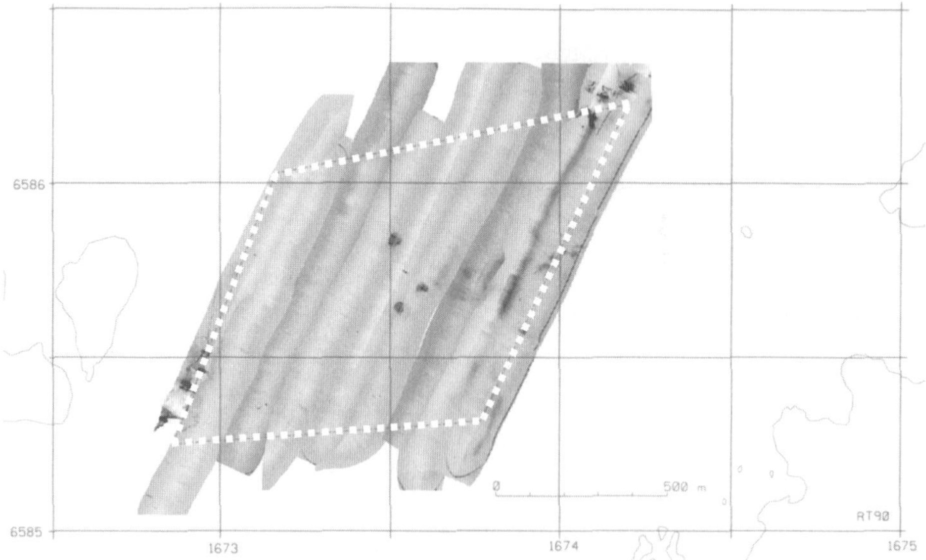

Figure 11.2. Traditional side-scan sonar mosaic covering the whole dumpsite. In the high-resolution picture over 450 objects were visible on the seabed.

Figure 11.3. The two ships used during the SITAR main sea trial: R/V *Altair* (top) and HMS *Fårösund* (bottom).

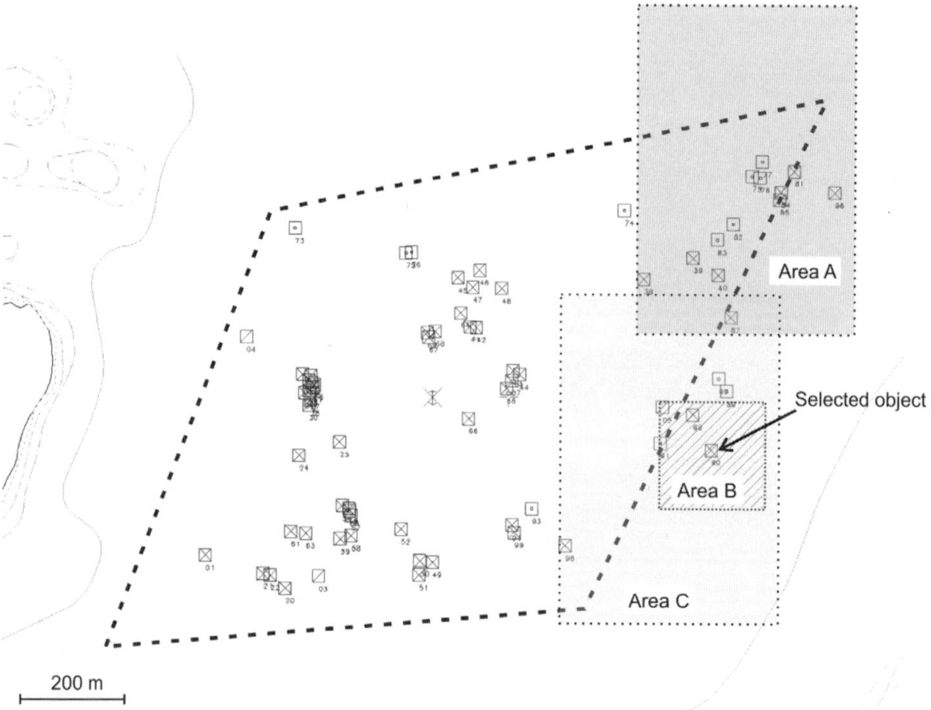

Figure 11.4. Dumpsite at Möja Söderfjärd. The area within the diamond-shaped figure corresponds to the one of identical shape of Figure 11.2. The acoustic trial covered the areas A, B and C. Marked squares are the positions of video inspections. "Selected object" marks the semi-buried object selected as the target in the multiple-aspect scattering measurements (Figure 11.6) performed in area B.

from FOI, KTH, KDA, ARMINES/EN, ISME, Bath University and NTNU participated at various stages of the experimental activities, under coordination from FOI.

The parametric side-scan sonar trials covered the whole dumpsite area, even if most tests were concentrated in the shallowest eastern region – A and C in Figures 11.4 and 11.5. In the selected region the water depth was around 70 m and – compared with the deeper part of the area, typically 100 m – the gas content in the sediment was slightly less, which ensured the best chance of acoustic penetration. The *Altair* made survey lines over the experimental area, and data were preliminarily analysed onboard to indicate the possible locations of dumped objects. ROV-operated video inspections of different bottom structures and potential targets as detected from the *Altair* were performed from HMS *Fårösund* for verification. Testing of multiple-aspect scattering measurement techniques were focused in area B where an identified semi-buried box was chosen as the test reference object (Figure 11.6).

The ROV was positioned within the area B by the use of a long-base ROV-Track positioning system. To start with, the seabed in the region for the target was densely

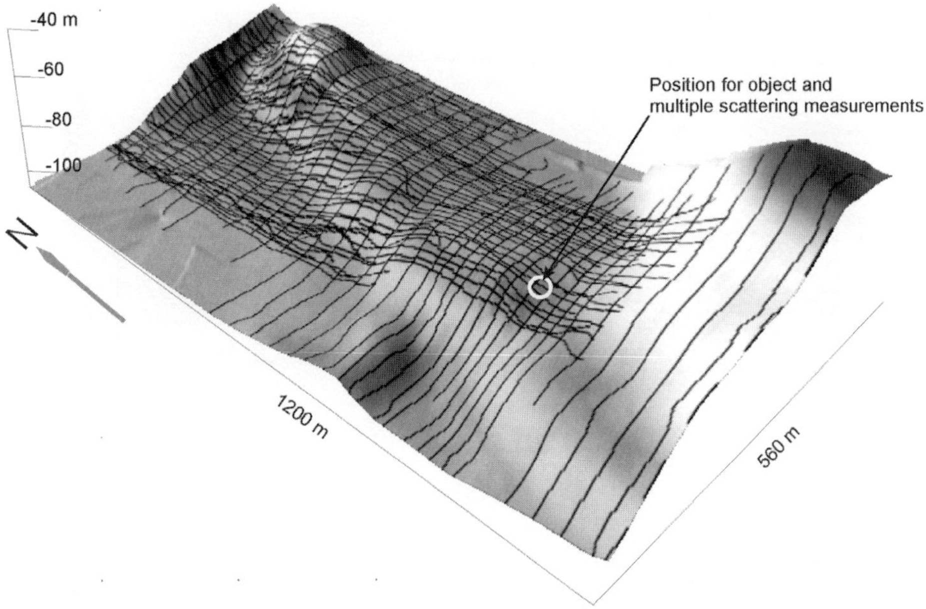

Figure 11.5. Bathymetry in the selected site (area A and area C); water depth was around 70 m, compared with 100 m in the main part of the dumpsite. The map also includes the extension of survey lines performed using a traditional sub-bottom profiler.

mapped through several crossing lines, whose schematic is illustrated in Figure 11.7. A standard (commercially available) parametric sonar was mounted in vertical mono-static mode on the ROV and the operating depth of typically 15–20 m above the bottom provided a small sonar footprint and high resolution. The purpose was to find the exact coordinates of the object within the ROV-Track long-base system, necessary information for the following multiple scattering measurements.

11.2 PARAMETRIC SIDE-SCAN SONAR (PSSS) OPERATION FROM THE R/V *ALTAIR*

The PSSS prototype was installed on the port side of the R/V *Altair* (Figure 11.8). Sonar tilt with respect to normal incidence had three pre-fixed values: 50°, 47° and 44°. The sonar was composed of three transmitting channels. Transmitter length (and thus SAS resolution) could be controlled by using one, two or three channels. Two lines of eight receiving channels were used. For every sub-array, two channels (HF and LF) were constructed. In addition, a DGPS and a motion reference unit (MRU) were used in order to provide, for each ping, the position and navigation information at the time of transmission. The overall system block diagram is given in Figure 11.9. A specific convention based on the NETCDF data format was developed in order to be able to

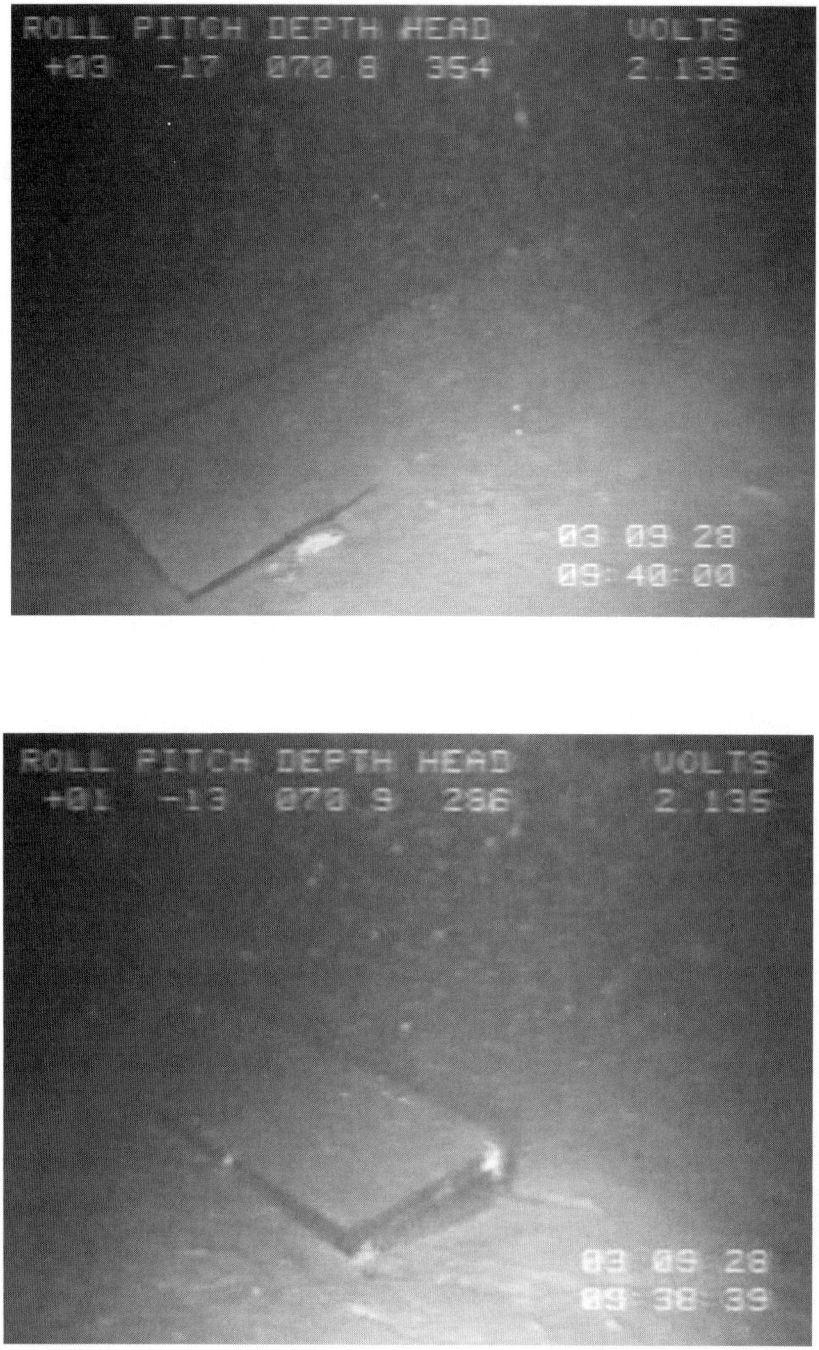

Figure 11.6. Two different views of the semi-buried box, the target for multiple-aspect measurements. The size of the box is approximately $1.3 \times 0.3 \times 0.3\,\mathrm{m}$.

Figure 11.7. The exact coordinates of objects within the ROV-Track long-base system, provided by four beacons, was found by mapping the area with several crossing lines. A bottom-penetrating, commercially available, parametric sonar was mounted in vertical monostatic mode on the ROV.

Figure 11.8. The parametric side-scan sonar prototype mounted on the frame is here ready for launch.

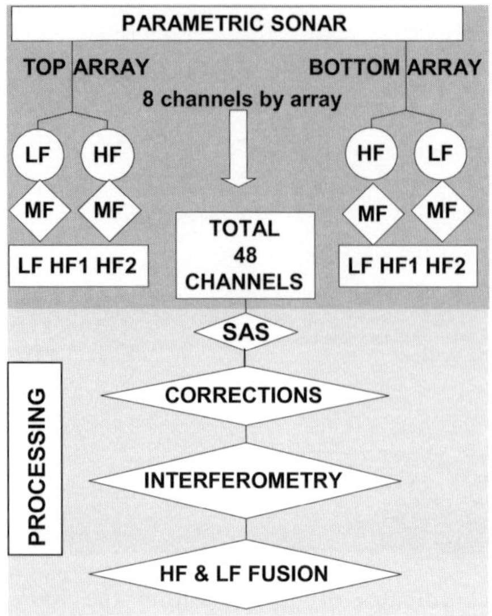

Figure 11.9. Block diagram of PSSS and processing.

handle and process the great amount of data recorded: 10 MByte/ping leading to about 1 GByte of raw data for an area of 100×100 m.

Experimental activities onboard the R/V *Altair* started on 22 September in the harbour at Slussen (Stockholm), with personnel from ISME, KTH, ARMINES/EN and KDA. On 25 September the *Altair* moved into position, and the PSSS became operational from 25 September to 2 October, with the sole exception of day 28, when strong wind and sea state conditions (sea state 4[2]) prevented operations. On 2 October a failure in the power unit of the system caused the anticipated end of the PSSS operation. Eighty-three lines were covered with various transmitted signals and tilt angles. About 500 GByte of raw data were stored on DVD and resulted in 1.5 terabytes of data after splitting the data into three sub-bands. The sum of all channels – in both high and low frequency – was displayed in real time by the ARMINES/EN-developed system, allowing an initial quality check of the data. In particular, the low-frequency data revealed themselves to be very noisy, possibly because of flow noise due to the frame installation of the system. A preliminary attempt onboard to filter the data showed that the noise could be suppressed at post-processing; however, it was jointly decided not to attempt to insert the filter in the acquisition and real-time display system, to avoid the risk of introducing bugs in the system without the necessary time to perform thorough testing. While this decision may be considered wise – since retrospective post-processing of the data allowed the suppression of low-frequency noise and no data were lost – it has also to be remarked that the unavailability of low-

[2] Sea state 4 is characterised by "small waves, becoming larger; fairly frequent white horses", according to the Beaufort Scale.

frequency data for inspection during the sea trial prevented the team from onboard detection of possible buried objects to which to direct HMS *Fårösund* for acoustic MAS and ground-truth checking.

11.3 MULTIPLE-ASPECT SCATTERING (MAS) MEASUREMENT SYSTEM

For the MAS experiment, an ROV-mounted single-beam parametric sonar (TOPAS 120, a commercially available system, not to be confused with the PSSS prototype developed by the SITAR team) was operated in several different positions around the target in combination with different pulse shapes and different emitter angles (see schematic in Figure 11.10).

The research ROV – *Plums* (platform for underwater measurement systems) – was upgraded to fulfil the requirements raised by the task (Figure 11.11). In particular, a new power supply, umbilical cable and network communication were installed. TOPAS was placed in front of the ROV using a mechanical joint, allowing tilting at different angles. In the configuration developed for the SITAR experiment, *Plums* was able to keep its attitude angles within an accuracy of 1°, and its depth within ±0.1 m. To track the horizontal position the ROV used an acoustic positioning system, ROV-Track, consisting of four transponders on the seafloor and a transceiver on the ROV. The geographic coordinates of the transponders were acoustically

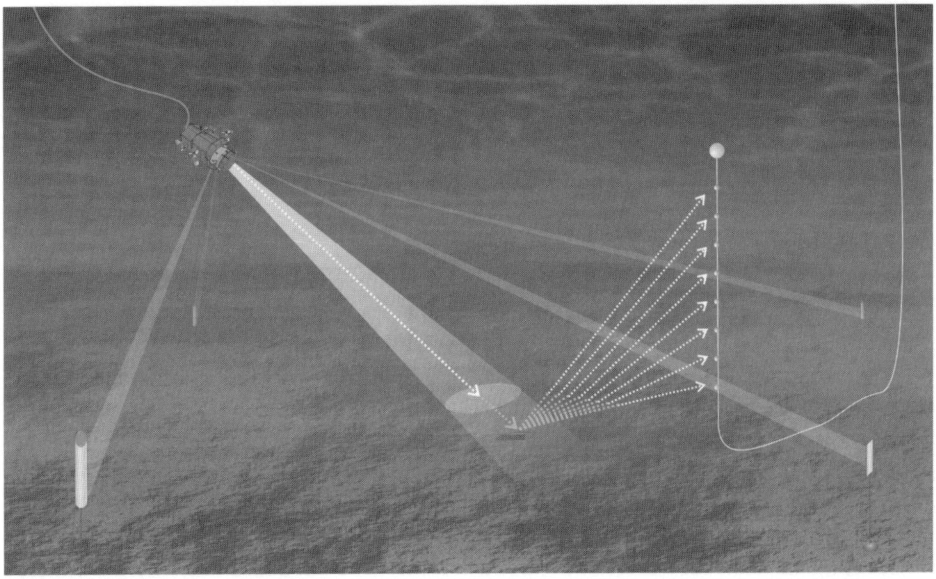

Figure 11.10. The multiple-aspect configuration where the transmitter and the receiving chain are separated around the target. Here, positioning of the ROV is provided by four acoustic beacons.

Figure 11.11. The ROV developed by FOI called *Plums* (platform for underwater measurement systems) equipped with the TOPAS 120 parametric sonar.

calibrated from HMS *Fårösund* with DGPS accuracy. A program was developed to synchronize the pings of the positioning system with the pings of the MAS sonar. This was necessary to avoid interference in both the sonar and the positioning systems.

The experimental system – consisting of an ROV with full freedom of movement and attitude in three dimensions, a narrow sonar beam, a target and a receiver chain – was quite complex. To facilitate navigation, a computer program was developed. The program read attitudes and depth from the ROV and horizontal position from the positioning system. The ROV, target(s), sonar beam, receiver array and other information was presented to the operator in several views on the screen (Figure 11.12). As an example, a screen enlargement showing the cross-section of the beam hitting the bottom is shown in Figure 11.13. Through the system, several operations can be commanded and executed automatically, like locking the beam onto a target, executing linear scans at a fixed transmitting angle, etc.. This was done by sending set-points for the pitch and course angles to the ROV's control electronics. In this way it was possible to keep the sonar beam locked on the target while the ROV was slowly moved around by the operator. The navigation program also facilitated the deployment and positioning of the receiver chain. All raw data were logged by the ROV control program and the positioning system. Data were also recorded by the navigation program. These data could later be replayed by the navigation program and presented in the same user interface.

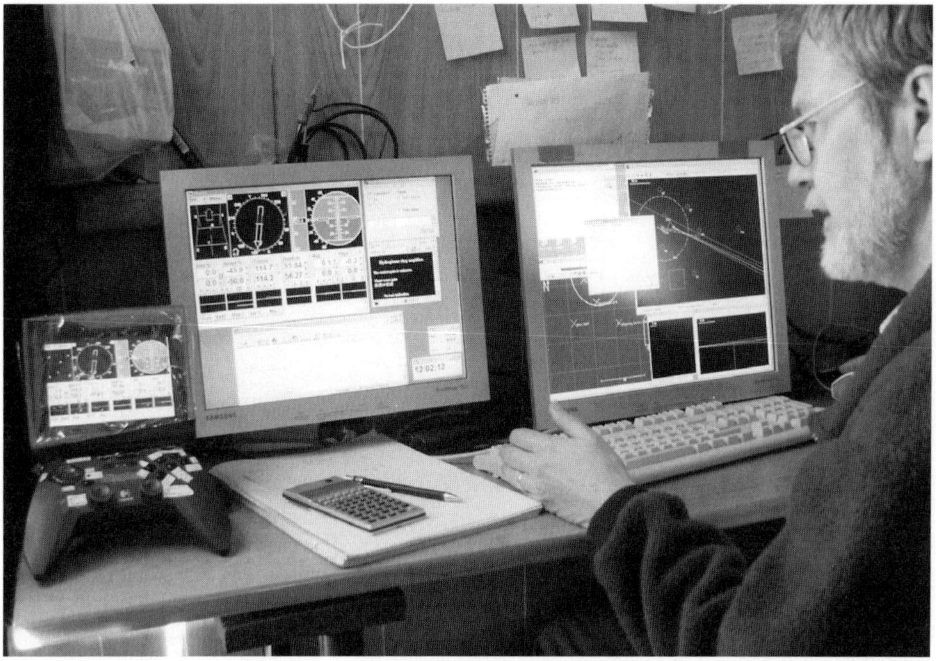

Figure 11.12. Interface designed to operate the ROV and parametric sonar for the multiple-aspect measurements.

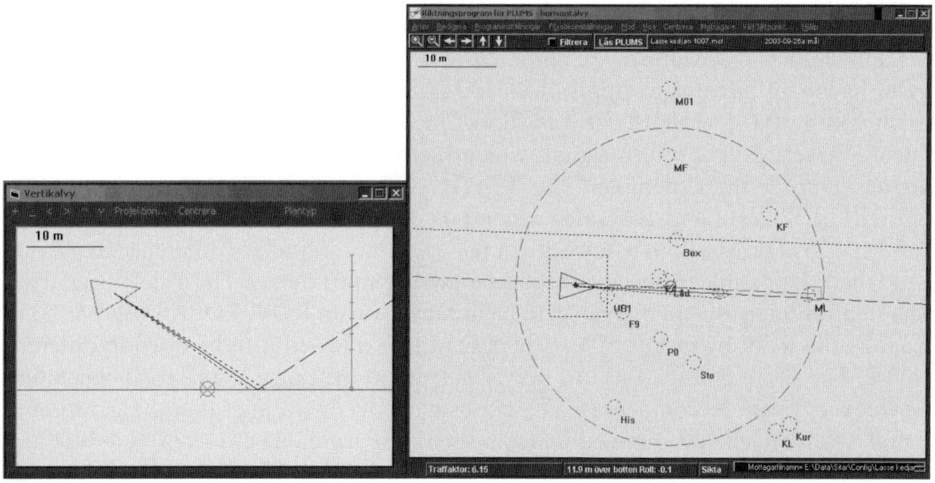

Figure 11.13. Example of some functions developed to operate and steer the ROV. The parametric sonar had a very narrow beam width (3–4°) which meant that the footprint was close to a meter in these measurements. For multiple-aspect measurements the accuracy of the positioning and steering were crucial.

Figure 11.14. The data acquisition system onboard HMS *Fårösund.*

The ROV was equipped with a parametric sonar, TOPAS 120, together with a six-element circular receiver array. Compared with conventional sources, the parametric sonar had many advantages appropriate to MAS application. The sonar had a narrow beam width of 3–4° – that is, it was almost independent of the frequency. Very low sidelobe levels gave a return signal, which had low reverberation and little volume scatter. During the experiment the TOPAS 120 transmitted primary frequencies centred on 120 kHz and generated secondary frequencies within the low-frequency band, 2 – 30 kHz. The high bandwidth in the operating secondary difference frequency band made it possible to generate broad-banded or short pulses. The transmitter array was small enough (40-cm diameter) to be mounted on an ROV. TOPAS was placed in front of the ROV by means of an attachment that allowed it to be tilted in different angles. The shortcoming of its low power output, in the range 165–190 dB depending on frequency, can be compensated by choosing short operating ranges. The reflected signal was received and summed up from the six hydrophone elements, resulting in a combined sensitivity typically inbetween −130 and −140 dB re 1 μPa @ 1 m in the calibrated frequency interval 2–18 kHz. The signals were amplified to 40 dB and distributed by cable to the data recording system (Figure 11.14), where amplification and bandpass-filtering were adjusted to pulse type and sampling rate.

At an early stage of the sea trial the primary receiving hydrophone chain suffered

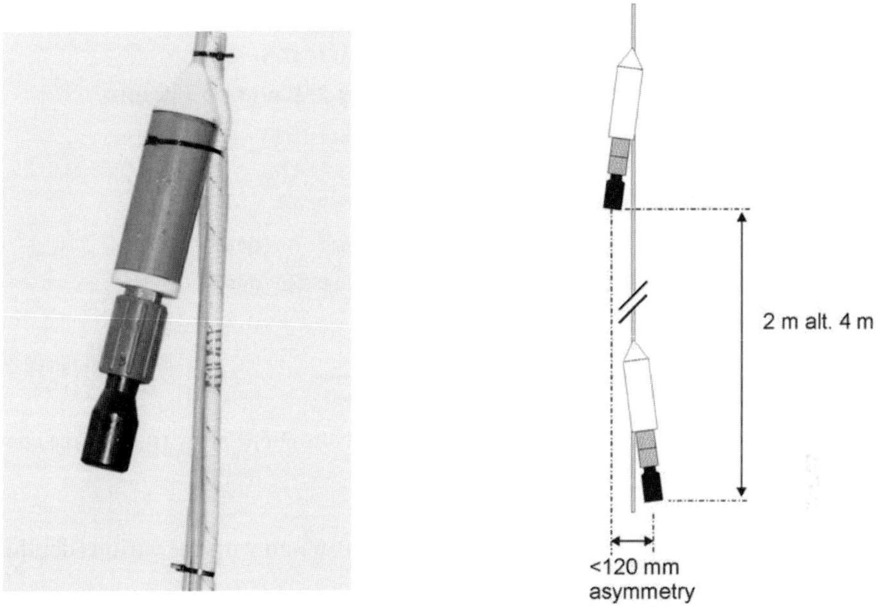

Figure 11.15. Vertical hydrophone chain geometry illustrated by the outline of two elements, and a picture of a mounted hydrophone.

irreparable damage. A spare hydrophone chain had to be used in the multiple aspect measurements, causing the receiving system to be different from the one originally planned. In particular, the used hydrophone chain consisted of five individual elements mounted along a line. Scattering measurements were performed using two different lengths of the chain, 10 and 8 m, respectively. The spacing between every two hydrophones was fixed at 2 m. In the 10-m-long chain, the spacing of the last pair of hydrophones was 4 m. The mounting of the chain caused a certain degree of horizontal asymmetry, up to 120 mm, between each hydrophone, which had to be taken into account in the analysis (see Figure 11.15). The receiving array was deployed in a vertical position and arranged with the deepest hydrophone 9 m above the bottom. The hydrophones were omnidirectional broadband hydrophones with a sensitivity of -160 dB (± 5 dB) re 1 μPa at 1 m within the calibrated frequency range 5–30 kHz. Each hydrophone included a low-pass filter at 30 kHz to avoid any disturbance from the transmitting primary frequencies. In Figure 11.16 the intended and the actual array configurations are shown.

To find the exact coordinates of the object, within the deployed long-base system, the seabed in the region for the target was mapped by the ROV with the TOPAS sonar in vertical monostatic mode with several crossing survey lines, resulting in the precise location of the object (Figure 11.17, see colour section). Knowing the position of the scattering object, the receiver hydrophone chain was deployed for multiple scattering measurements. Measurements were performed with the receiving vertical hydrophone chain located in three different positions around the object. The horizontal distances

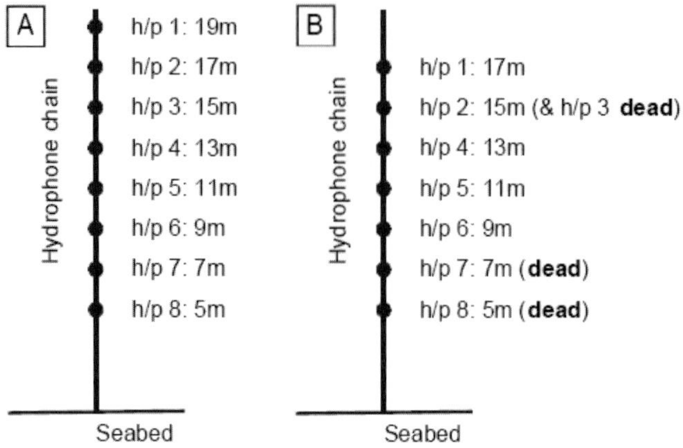

Figure 11.16. Failed hydrophone array (A) and spare hydrophone array (B) with the configuration finally used in the MAS experiment.

between object and chain varied from 10 m up to 20 m and with the various depths of the hydrophones.

For each chain position, the ROV was operated in several different positions around the target in combination with different angles of the parametric sonar. Oceanographic measurements (conductivity, temperature and sound speed profiles) were conducted periodically throughout the whole trial. The most evident variations of the measured parameters were in the upper part of the water column. As an example, the water sound speed close to the surface decreased from 1463 m/s at

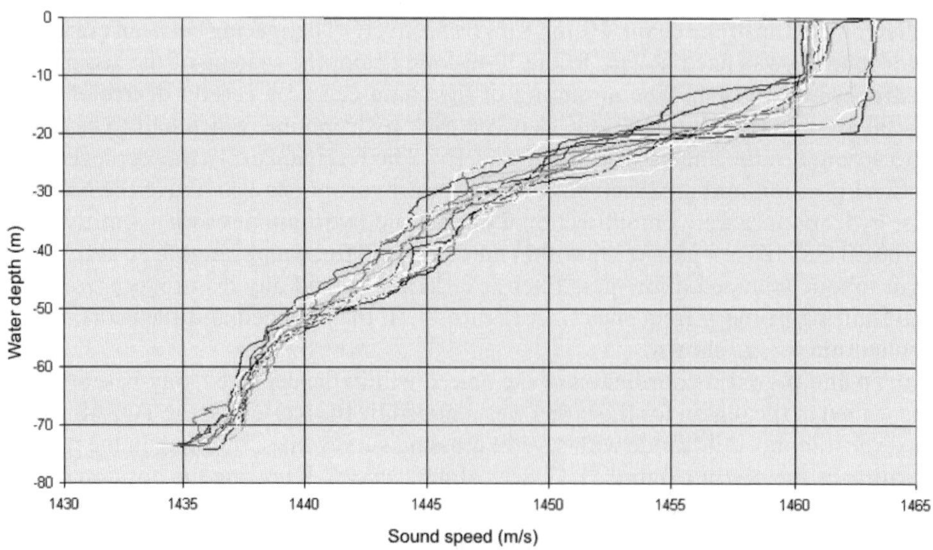

Figure 11.18. Measured variations in water sound speed during the period of the sea trial.

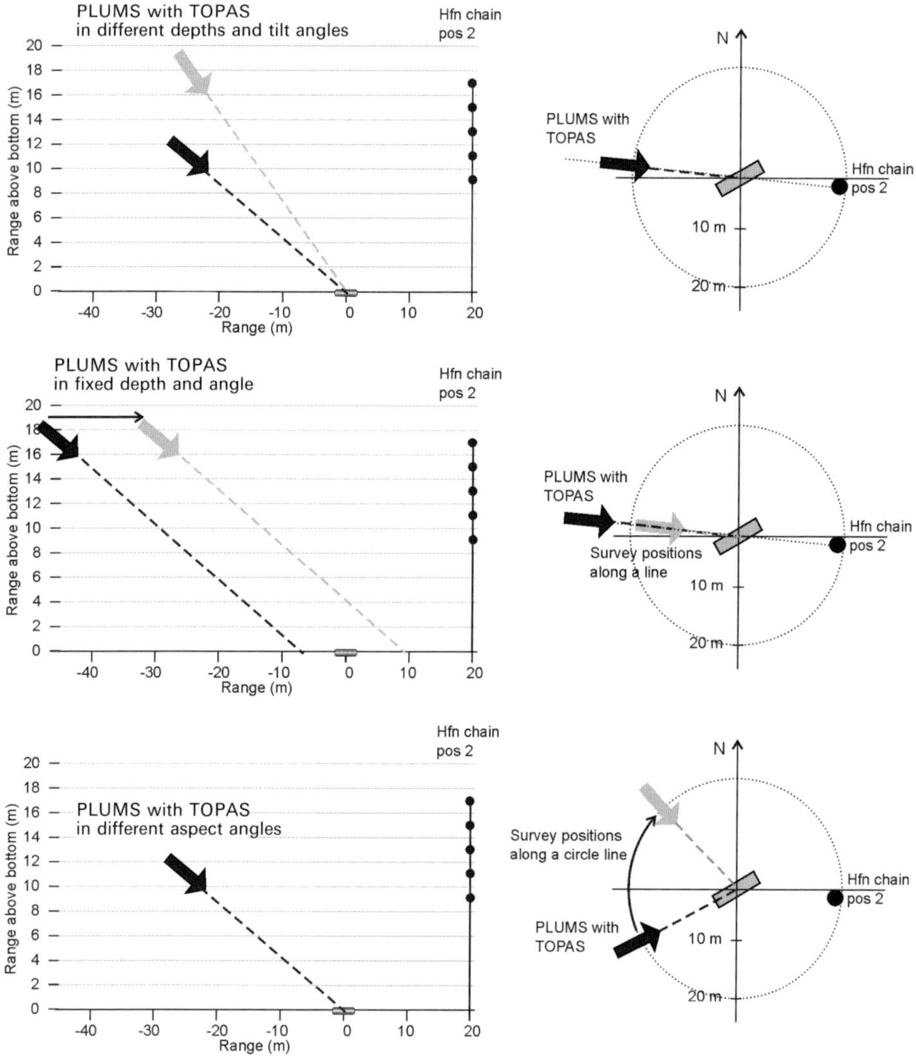

Figure 11.19. Three examples of multiple-aspect survey geometries tested during the trial.

the beginning of the trial to 1457 m/s at the end. The sound speed in the depth region of the multiple scatter surveys, up to 20 m above the bottom, was more stable during the experiment period and the variations much smaller (Figure 11.18). Examples of three different multiple-aspect survey methods are given in Figure 11.19. Five different pulses were used, Ricker[3] pulses (5 kHz, 10 kHz and 20 kHz) and chirp pulses (8–16 kHz and 10–20 kHz).

[3] A Ricker signal is a specific, pulse-like, signal shape very often used in marine seismic exploration.

Part IV

Results

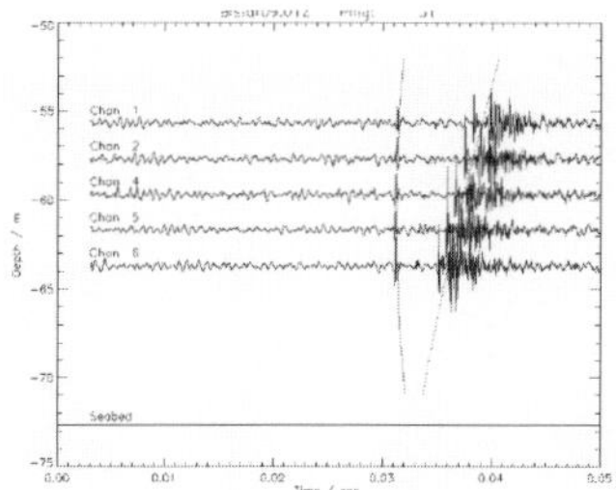

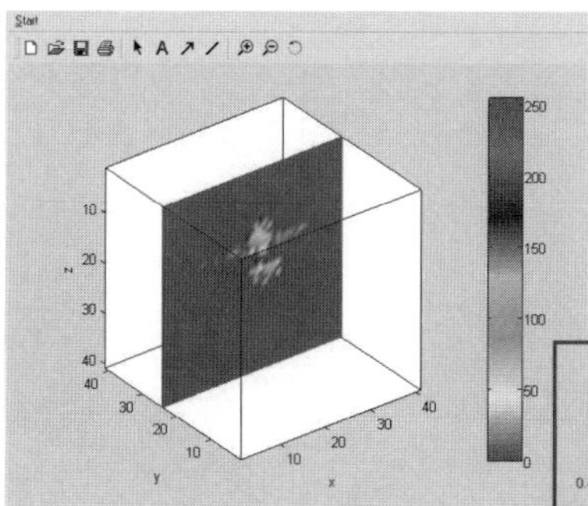

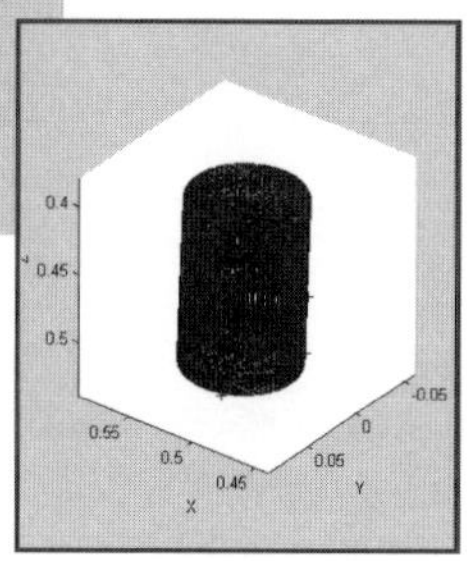

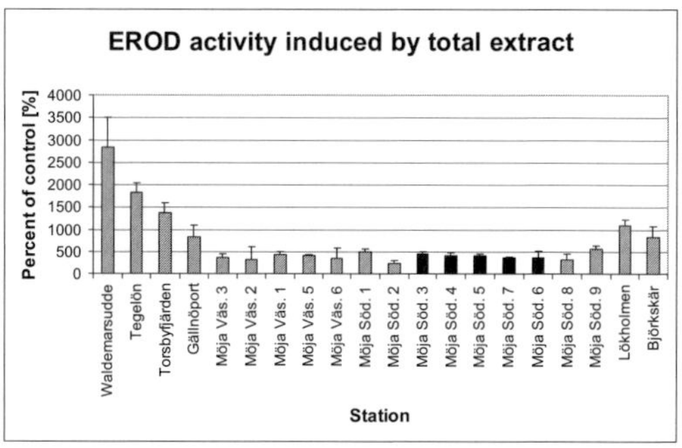

12

Geoacoustic inversion

M. A. Larsen and J. M. Hovem

12.1 INTRODUCTION AND BACKGROUND

Geoaoustic inversion is the process of estimating sea bottom parameters from geoacoustic modelling based on measurements. One of the classical references on this topic is that of Hamilton (1980), which has the following definition of geoacoustic modelling: a "geoacoustic model is a model of the real sea floor with emphasis on measured, extrapolated, and predicted values of those properties important in underwater acoustics and those aspects of geophysics involving sound transmission." The product of sound speed c and density ρ is the acoustic impedance, a measure of how "hard" the sea bottom is. An estimate of acoustic impedance in the upper part of the bottom would be useful, as it tells us how much of the transmitted energy is absorbed in the bottom. Since the bottom surface is more or less irregular, an estimate of roughness would also be valuable in characterizing the bottom.

As part of the experimental activities of the SITAR project in the Stockholm Archipelago, fall 2003, an acoustic source was placed on an ROV to image the seabed, and the scattered field was measured on several hydrophones; the acoustic source (the TOPAS 120 parametric sonar system, see Chapter 11) was used in much the same way as an echosounder, transmitting acoustic pulses with centre frequencies of 5 kHz, 10 kHz and 20 kHz at normal incidence with respect to the bottom. Several transects were run in this configuration over an area comprising the semi-buried barrel chosen as the target for investigation. The acoustic data gathered from these runs have been processed in order to obtain a geoacoustic model of the upper sediment strata at the experimental site.

To obtain bottom parameters from the measured data, a method named FARIM (Berntsen *et al.*, 1999) – short for "frequency analysis based roughness and impedance estimation method" – was used. An estimate of the acoustic impedance from calibrated data and an estimate of the roughness of the sea bottom were obtained. The heart of the method involved expressing the coherent field scattered in the specular direction. This expression is based on the observation that a rough bottom scatters energy incoherently away from the receiver. Calculating the bottom impedance simply from the ratio of received-to-transmitted energy at the transmitter/receiver location (corrected for geometric spreading loss in the water), therefore, generally results in underestimation of impedance. An estimate of roughness was calculated by comparing the power spectrum of the received signal and the power spectrum that would be received according to the theory by expressing coherent scattering in the specular direction. An improved estimate of bottom impedance was then obtained by correcting for incoherent scattering caused by this roughness. Although the FARIM method was developed for characterizing bottom sediments, we will also see that it is useful for detection of buried objects in bottom sediments.

In Section 12.2 we discuss monostatic measurements in the main sea trial; the theory of scattering from rough surfaces is then briefly reviewed, the FARIM method illustrated, results from analysing the data shown and finally a critical discussion of the data processing methods used and of the results obtained is given.

12.2 THE MAIN SEA TRIAL AND MONOSTATIC MEASUREMENTS

The *Plums* ROV was used with a TOPAS 120 parametric echosounder with half-power beam width of approximately ±3 degrees (see Chapter 11). The transmitting antenna could be tilted at an angle relative to the ROV. A transceiver on the ROV and four transponders deployed on the seabed comprised an LBL (long-baseline) acoustic positioning system, and the geographical position of the ROV could be tracked with repeatability of a fraction of a metre. Measurements of the pitch and roll of the ROV were taken and corrected-for in processing. The pulse repetition rate was 300 ms, and the recorded time for each reception (ping) was 20 ms. The exact time of each ping was measured using a GPS clock. The received signals were amplified and bandpass-filtered. Data were digitized at a 200-kHz sampling rate with 16-bit precision and stored on disk. Short transients (approximate Ricker[1] pulses) were transmitted into the bottom. Different frequency contents were used in the pulses. The lower frequencies penetrated the bottom while higher frequencies gave better resolution (Widess, 1973). The shapes of the transmitted Ricker pulses measured using a reference hydrophone array are shown in Figure 12.1 (see colour section). Monostatic data were measured by the ROV running over the target and its immediate bottom area. The 5-kHz data are shown in Figure 12.2 (see colour section), where the ROV is about 17 m above the bottom, and left to right represents the south to north direction. The 10-kHz data are shown in Figure 12.3 (see colour section), and the 20-kHz data are shown in Figure 12.4 (see colour section). The ROV is about 11 m above the bottom in both data sets, and left to right in these figures represents the northeast to southwest direction. The bottom depth in the area is around 73 m. The target was visible in the

[1] A Ricker signal is a specific, pulse-like, signal shape very often used in marine seismic exploration.

data as a strong amplitude anomaly. The approximate distance travelled is shown, to get an idea of the size of the object.

12.3 SCATTERING FROM ROUGH SURFACES

The laws of reflection of a plane wave by a plane surface are well understood. Imagine that the bottom surface is plane, and water with density ρ_w and sound speed c_w is overlying a homogenous bottom with density ρ and sound speed c. The pressure–amplitude reflection coefficient $\Re$ at the plane surface separating the two fluid media (we will ignore shear wave conversion in the bottom), for a plane wave at grazing incidence θ in the water is:

$$\Re = \frac{\rho c/\sqrt{(1 - c^2 \cos^2 \theta / c_w^2)} - \rho_w c_w / \sin \theta}{\rho c/\sqrt{(1 - c^2 \cos^2 \theta / c_w^2)} + \rho_w c_w / \sin \theta}. \tag{12.1}$$

However, physical surfaces such as the sea bottom are generally not planar. When the boundary is irregular, with periodic or random variations of height measured from a certain mean level, the laws of reflection are not known in exact form. Perhaps the most striking difference in the behaviour of a smooth surface and a rough surface is the fact that a smooth surface will reflect the incident wave specularly in a single direction, whilst a rough surface will scatter it in various directions. This scattered energy is usually termed the diffuse or incoherent field, because of its angular spread and lack of phase relationship with the incident wave. Similarly, the specular wave is often termed the coherent field, due to its predictable, and constant, phase relative to that of the incident wave.

The angle of a particular frequency component (phase) ϕ in any direction is determined by height variations across the surface and cannot be predicted without knowledge of the surface profile. To obtain a quantitative formulation of scattering from rough surfaces, we will therefore use statistical methods to describe surface irregularities. The deviation of height from a mean level is assumed to be a random variable with a Gaussian probability density. The assumption of a Gaussian probability density is probably the most general assumption, and leads to a fairly simple expression for the specular field. The coherent field (averaged phase) for a particular frequency component f is given as

$$\langle \exp(i\phi) \rangle = \exp[-2(\sigma \sin \theta 2\pi f / c_w)^2] \tag{12.2}$$

with σ being the RMS roughness. Angular brackets $\langle\ \rangle$ indicate an ensemble average over surface realizations. One of the first publications of this important result was Eckart (1953). Diffuse fields are excluded, as these will be of random phase and sum to zero. Equation (12.2) can be interpreted as a reflection coefficient for a *rigid* rough bottom. It illustrates the important point that the roughness of any scattering surface is not an intrinsic property of the surface itself but depends on the properties of the waves being scattered. Both the frequency and direction of the incident wave determine how rough the surface appears to be. Since a surface will reflect specularly if

$\sigma/\lambda \to 0$, where $\lambda = c_w/f$ is the wavelength, the result is valid for a *slightly rough* surface but not for a *very rough* surface. A thorough discussion of scattering from rough surfaces can be found in Beckman (1987). For a rough surface the local angle of incidence θ in Equation (12.2) is not constant, but averaging yields the overall angle of incidence, which is constant for the whole surface. For a plane wave at normal incidence, we will therefore set $\theta = \pi/2$. We can thus determine the acoustic impedance of the bottom from

$$Z = \rho_w c_w \frac{1 + \Re}{1 - \Re}, \tag{12.3}$$

if we can obtain an estimate of the reflection coefficient.

12.4 THE FARIM METHOD

The FARIM method was developed in the European ISACS project (Berntsen *et al.*, 2000) – short for "integrated system for analysis and characterization of the seafloor." The purpose of the method is to obtain an estimate of the acoustic impedance and RMS roughness from calibrated measured data. The experimental data configuration for the FARIM method exactly matches that described previously as part of the activities in the SITAR field trial. In particular, the transmitted short pulses contain a band of frequencies and, since we want to deal with time-averaged quantities, we will (Fourier-) transform the signals to frequency domain representations. We can calculate an *energy reflection coefficient* as

$$R = \frac{\int_f P_r(f)\,df}{\int_f P_i(f)\,df} \tag{12.4}$$

where $P_r(f)$ and $P_i(f)$ are power-spectral densities of the *reflected* and the *incident* signal, respectively, close to the bottom. What we have available are calibrated values of the received and the transmitted signal at the transmitter/receiver at some distance d above the bottom. These energy levels are thus corrected-for with $\pm 20\log(d)$, using a spherical spreading law, to obtain their values at the bottom surface. Equation (12.4) inserted into (12.3) generally results in underestimation of impedance since energy is scattered incoherently away from the receiver. From (12.2) we define a *coherent energy reflection coefficient* for normal incidence:

$$R_{coh}(f;\sigma) = \exp[-4(\sigma 2\pi f/c_w)^2]. \tag{12.5}$$

It is seen in Figure 12.5 (see colour section) that a rough bottom acts effectively as a low-pass filter on the transmitted pulse. The signal spectrum that would be received according to theory is calculated and compared with the actual received signal spectrum. The spectral shapes are captured by the *centre-of-gravities* of the spectral densities, and it is this scalar parameter that is computed and compared to estimate

RMS roughness:

$$\hat{\sigma} = \underset{\sigma}{\operatorname{argmin}} \left| \frac{\int_f f R_{coh}(f;\sigma) P_i(f)\, df}{\int_f R_{coh}(f;\sigma) P_i(f)\, df} - \frac{\int_f f P_r(f)\, df}{\int_f P_r(f)\, df} \right| \tag{12.6}$$

The estimated σ inserted into the coherent energy reflection coefficient is used to find a modified energy reflection coefficient:

$$R(\hat{\sigma}) = \frac{\int_f P_r(f)\, df}{\int_f R_{coh}(f;\hat{\sigma}) P_i(f)\, df}. \tag{12.7}$$

Equation (12.7) inserted into (12.3) yields an improved acoustic impedance estimate of the bottom.

At this point it is useful to discuss the assumptions and approximations made in the FARIM method. By using (the square root of) R from (12.4) or (12.7) instead of $\mathfrak{R}$ from (12.1), when finding the impedance from (12.3), the surface boundary can be approximated at each point by the tangent plane. This approximation together with the far-field assumption that the wave is locally plane are commonly referred to as the *Kirchhoff approximation.*

The tangent–plane assumption will obviously be good when the radius of curvature of the irregularities is large compared with the wavelength of the signal, but will break down completely if the roughness includes sharp edges or sharp points. This suggests that vertical RMS roughness is not sufficient to describe the properties of an irregular surface: a horizontal roughness criterion is also needed. The correlation distance L_c is defined as the distance in which the normalized correlation of two arbitrary points on the surface drops to the value e^{-1}. $L_c = 0$ implies a surface that tends to be everywhere discontinuous, while a large L_c implies a gently rolling surface. For the Kirchhoff approximation to be accurate, we must require $L_c \gg \lambda$. A direct consequence of the constraints we have put on vertical roughness and surface correlations is that the mutual interactions of the irregularities (shadowing and multiple scattering) are ignored.

12.5 DATA ANALYSIS

We have analysed the 5-kHz, 10-kHz and 20-kHz data sets using the FARIM method. Figures 12.6, 12.7 and 12.8 (all in colour section) show FARIM analyses on the 5-kHz data. Figures 12.9, 12.10 and 12.11 (all in colour section) show FARIM analyses on the 10-kHz data. Figures 12.12, 12.13 and 12.14 (all in colour section) show FARIM analyses on the 20-kHz data. Figures 12.6, 12.9 and 12.12 show estimated impedance from Equations (12.4) and (12.3). Figures 12.7, 12.10 and 12.13 show estimated roughness from Equation (12.7). Figures 12.8, 12.11 and 12.14 show estimated impedance from Equations (12.7) and (12.3). The estimated values at each ping are plotted with colour-coded dots. This type of display illustrates a way of visualizing bottom parameters over a larger area. The scale of change of bottom properties can thus be observed. Impedance anomalies in other positions could be caused by other objects, which have been verified from camera inspections to exist in the area.

12.6 DISCUSSION

We have discussed the FARIM method for geoacoustic inversion of monostatic data from the SITAR project. The acoustic impedance of the upper part of the bottom was estimated. An estimate of roughness was obtained by comparing the signal spectrum that would be received according to theory with that actually received. The theoretical spectrum is a low-pass filtered version of the spectrum of the transmitted pulse. Correcting the bottom impedance estimate for the effect of roughness resulted in higher acoustic impedance. Data with short transients – such as Ricker pulses – were used in the experiment. The pulses had centre frequencies of 5 kHz, 10 kHz and 20 kHz. FARIM analysis was carried out on the measured data.

Sound speed in water was measured throughout the sea trial (see Chapter 11). The variation in water sound speed during one of the days is shown in Figure 12.15 (see colour section). The average value of water sound speed close to the sea bottom during the time of measurement was about 1440 m/s. With frequencies of 5 kHz, 10 kHz and 20 kHz, this resulted in a wavelength in water of 0.29 m, 0.14 m and 0.07 m, repectively. As previously explained, the bottom surface will reflect specularly if the wavelength components of the transmitted pulse are very large compared with the roughness. Consequently, Equation (12.2) is valid if this condition is met. This implies that roughness calculation using the FARIM method will be more trustworthy when the transmitted pulse contains higher frequency components.

An attempt can be made to determine sound speed and density from acoustic impedance estimates of the sea bottom. Laboratory values of sound speed and density for different sediment types were taken from table 1B in Hamilton (1980). We have assumed that the underlying law relating sound speed to density is linear. With this assumption, the acoustic impedance is quadratic in sound speed or density, from the relation $Z = c\rho$. The parameters of the regression equation are found by minimizing least-squares errors. The sound speed and density can thus be determined from Figure 12.16 (see colour section) from the estimate of acoustic impedance.

For a parametric source like the TOPAS 120, the non-linearity of high-amplitude sound propagation was exploited to generate low-frequency sound. The source transmitted two high-frequency (primary frequency) signals that interacted in the water to give the desired difference frequency (secondary frequency). During data analysis using the FARIM method we assumed that the geometrical transmission loss in water obeyed a spherical spreading law. That assumption was indeed incorrect; the true spreading loss was a more complicated function of distance. A series of calibration measurements should be done to find the sound level as a function of distance from the source. The estimated values of sea bottom impedance found with the

FARIM method can thus be slightly erroneous. However, the height above bottom of the ROV with the parametric source was rather stable, and the estimated impedance values from one ping to another were therefore assumed to agree adequately relative to each other.

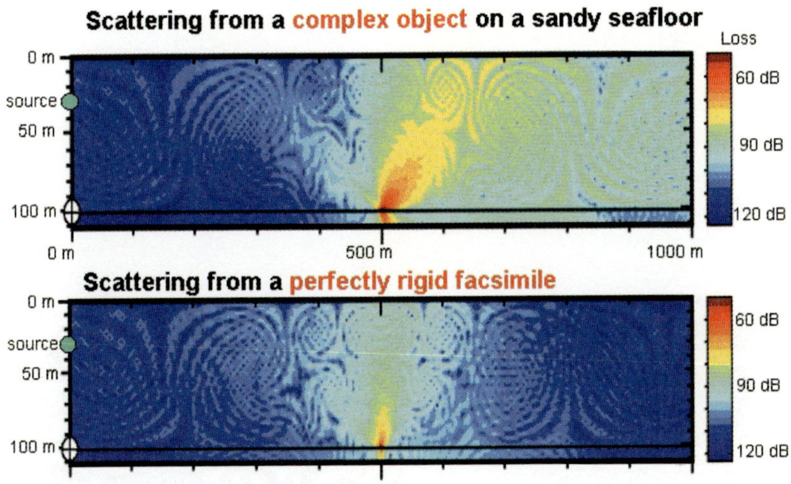

Figure 3.1. Acoustic scattering from apparently similar objects on a sandy seafloor, simulated using FESTA/OASES (image adapted from Pace, 2004). The black line shows the seabed.

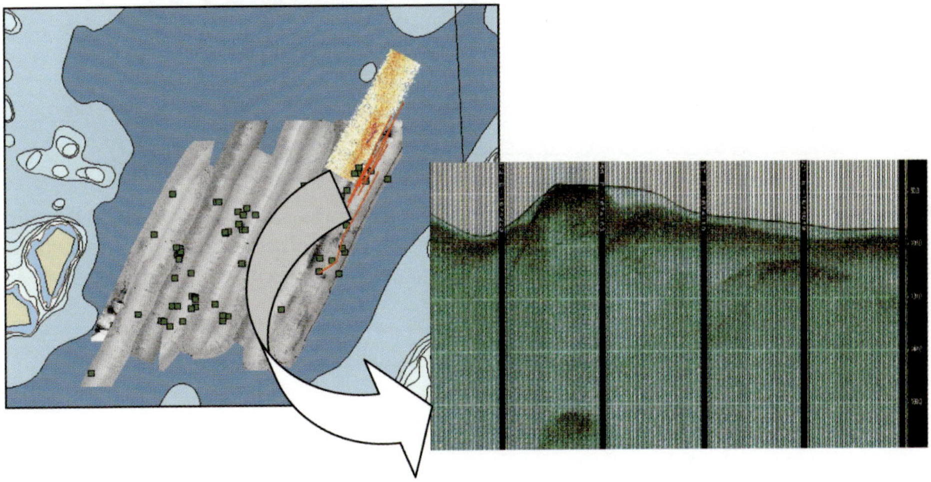

Figure 7.4. Thematic map simultaneously displaying bathymetry, object distribution, side-scan sonar (standard side-scan in grey levels, SITAR-developed parametric side-scan data in colour) and sub-bottom profile lines in red. Interrogation of a sub-bottom profiling line displays the raw data in a separate window.

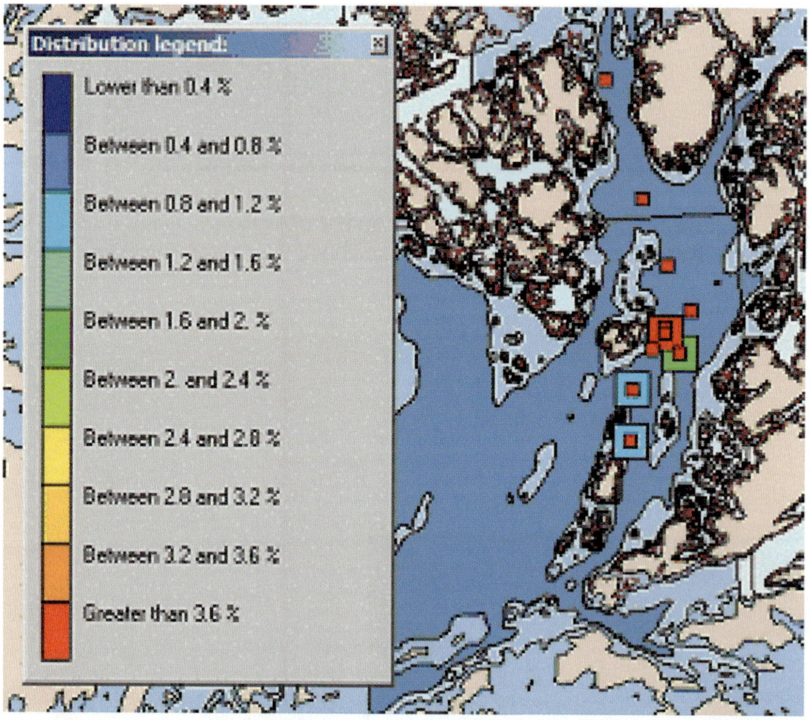

Figure 7.5. Thematic map of relative EROD-induced activity as measured from SITAR study site samples. The map is shown as an indication of the capabilities of the data presentation system – environmental implications of the map presented are discussed in Part IV.

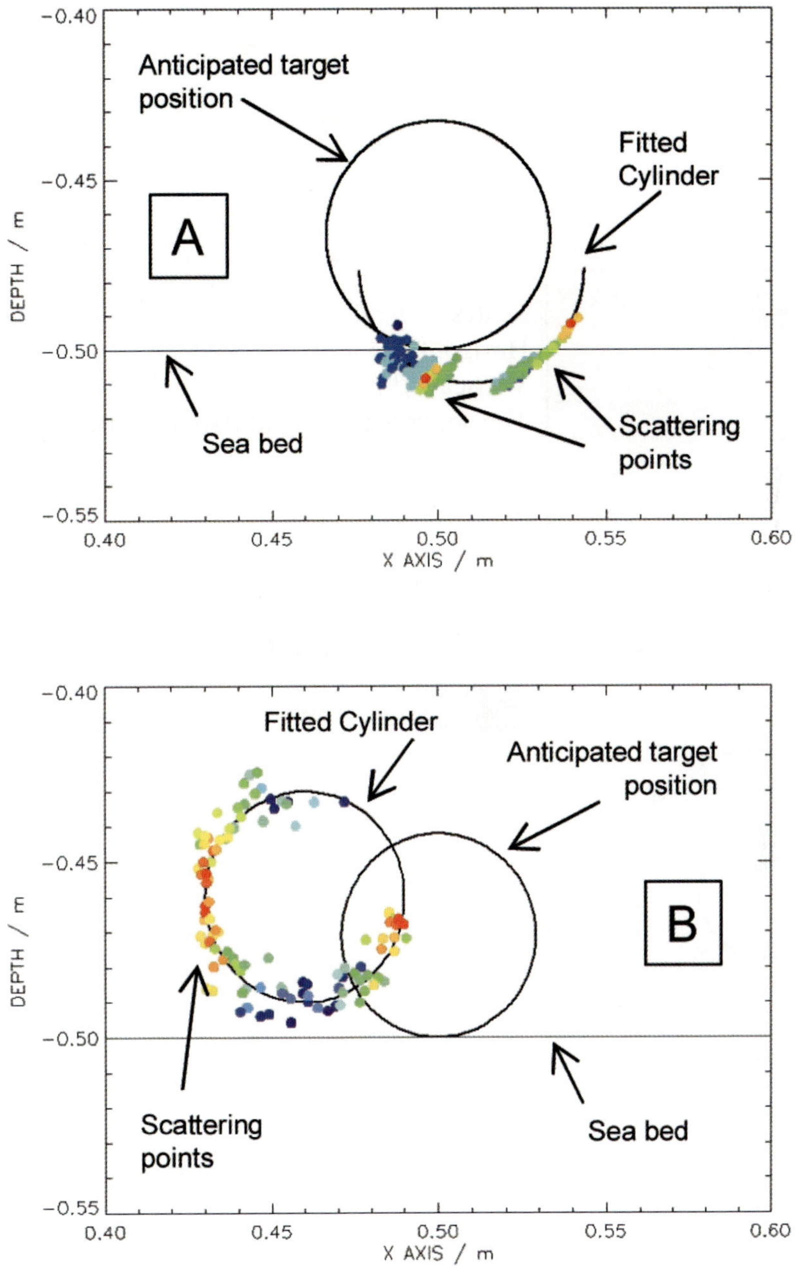

Figure 8.4. The scattering points, detected and localized using the algorithms developed, are colour-coded according to their strengths (from blue, low, to red, high). Comparison with the anticipated target positions for targets T_1 (A) and T_2 (B). The offsets from the expected positions are very small, and can be explained by the uncertainty in positioning the targets by hand from the top of the water-filled tank.

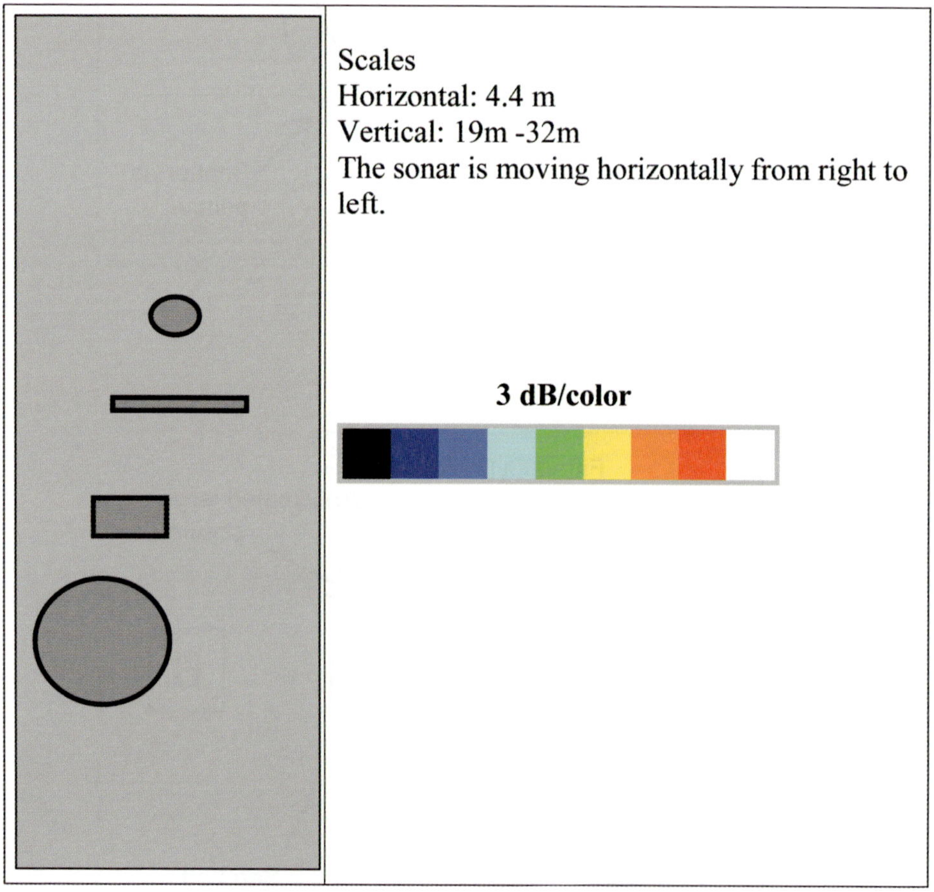

Figure 9.5. Scale of the tank experiment data.

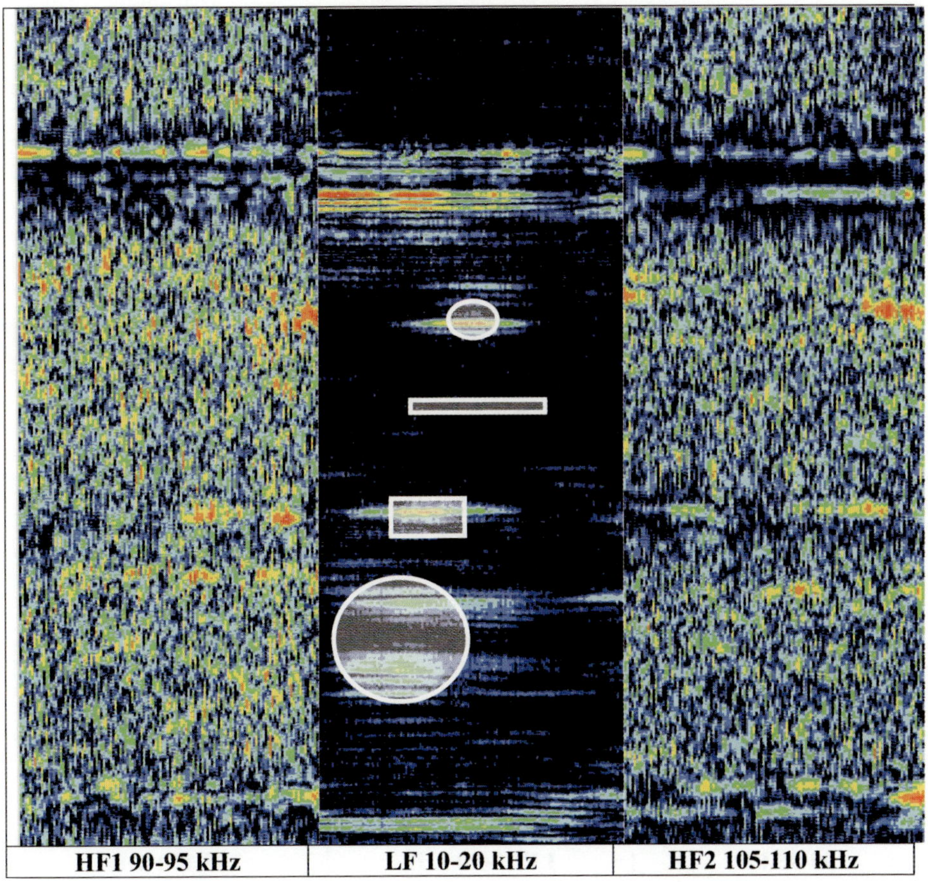

Figure 9.6. Raw data, imaging of the tank and buried objects.

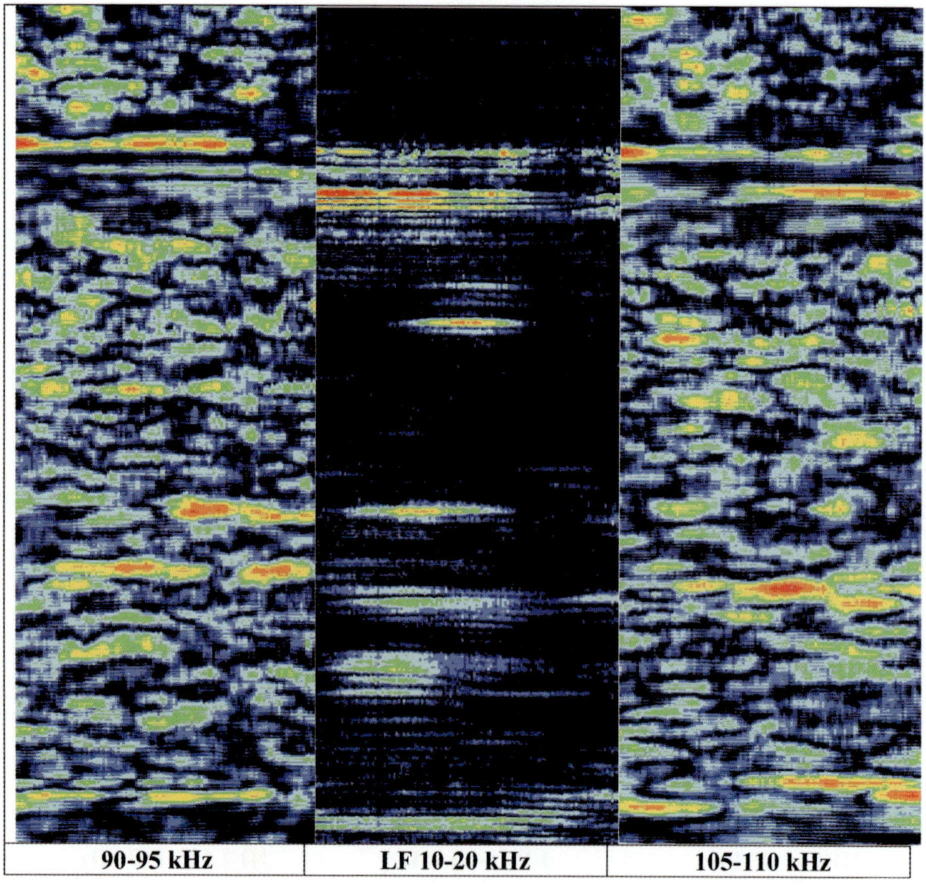

Figure 9.7. SAS processed data (of Figure 9.6).

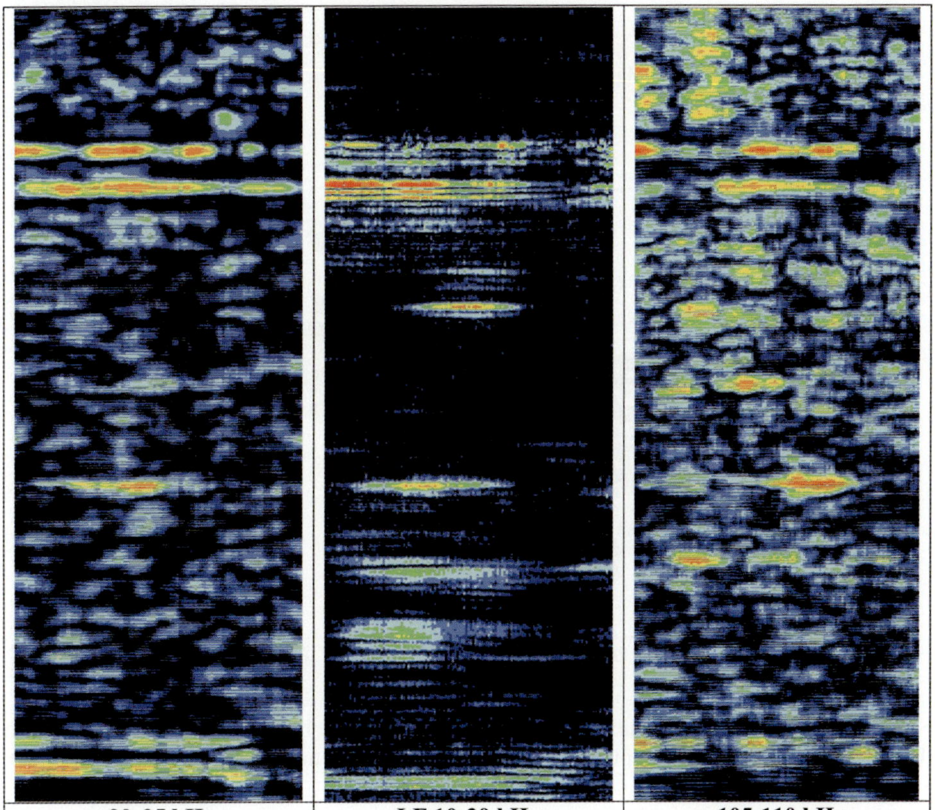

Figure 9.8. SAS-processed data + DPC autofocusing (of Figure 9.6).

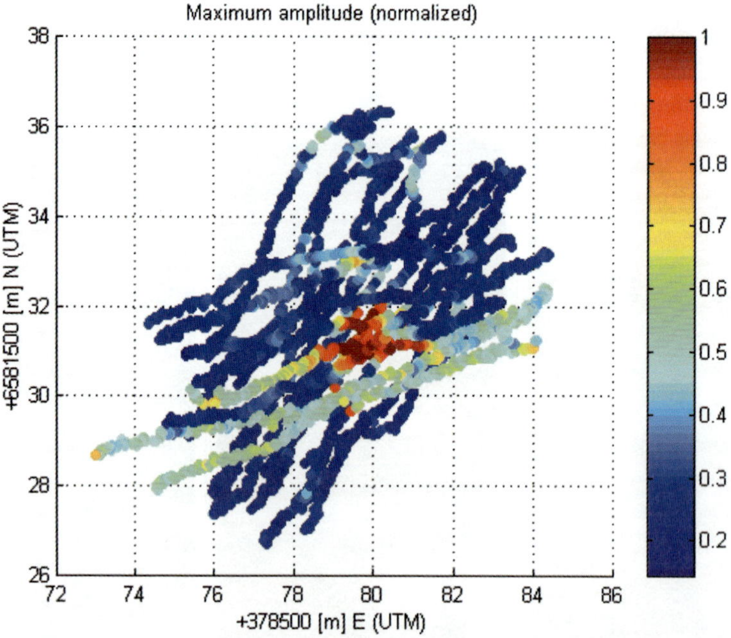

Figure 11.17. Survey lines in the area around the box of Figure 11.2. The lines are overlaid with the amplitude of the reflected pulse, indicating the exact position of the selected object (max amplitude).

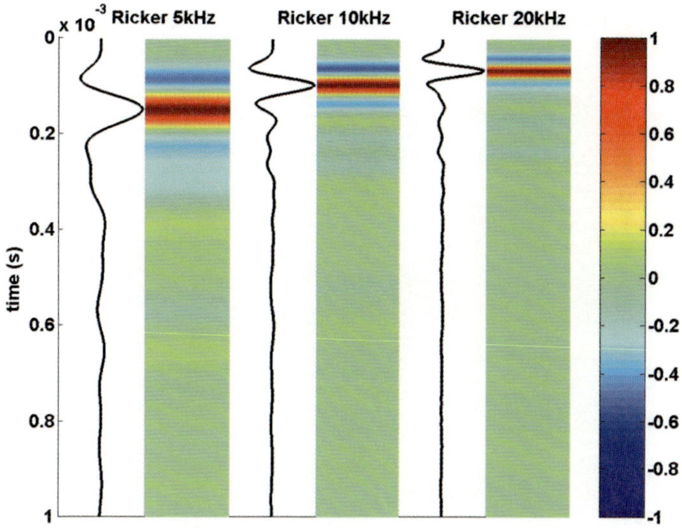

Figure 12.1. Transmitted pulses, Ricker 5 kHz, 10 kHz and 20 kHz.

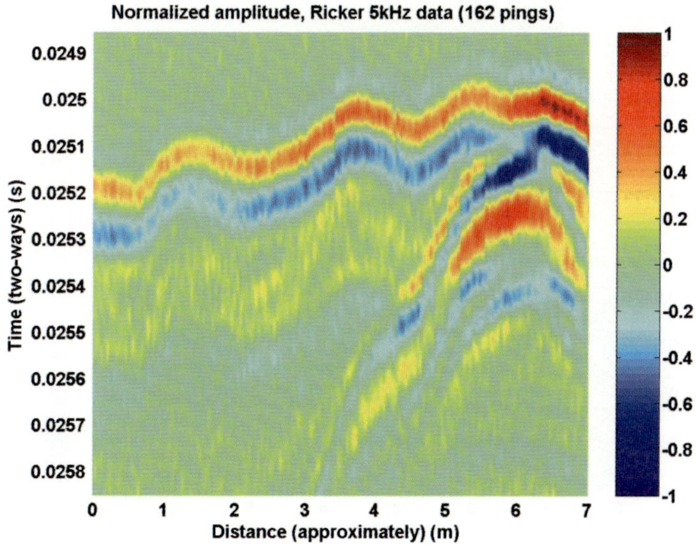

Figure 12.2. 5-kHz monostatic data. The target is to the right in the figure. The ROV is 17 m over the bottom surface.

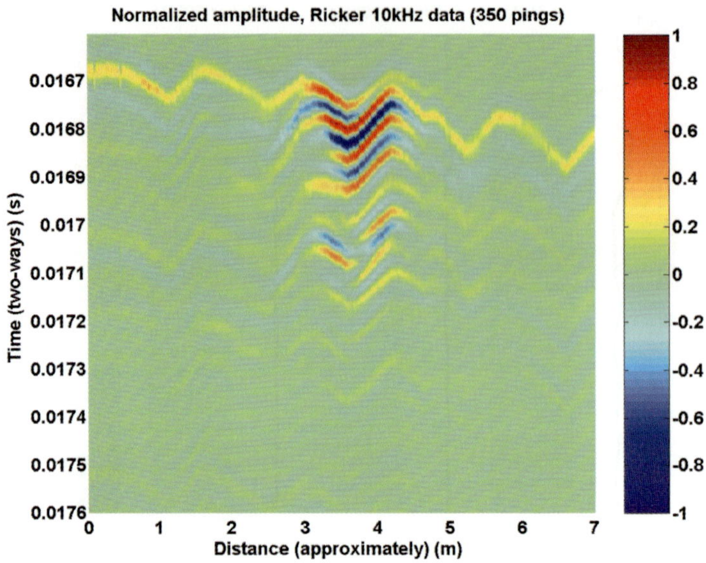

Figure 12.3. 10-kHz monostatic data. The target is in the centre of the figure. The ROV is 11 m over the bottom surface.

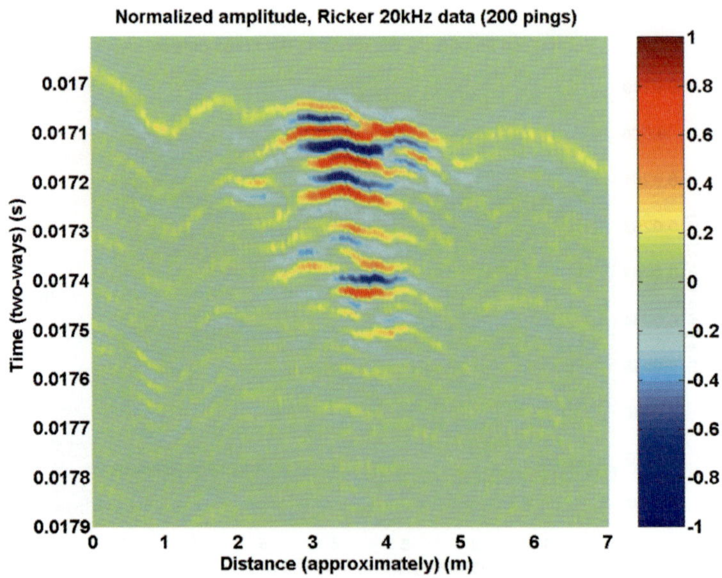

Figure 12.4. 20-kHz monostatic data. The target is in the centre of the figure. The ROV is 11 m over the bottom surface.

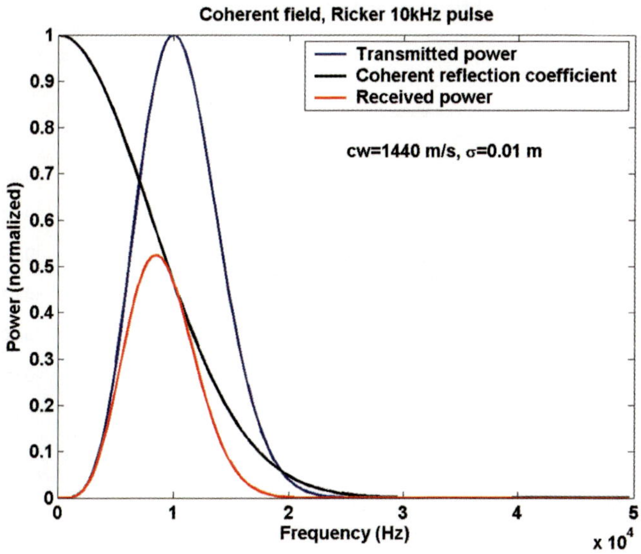

Figure 12.5. Coherently scattered energy. Power spectra for a 10-kHz theoretical Ricker pulse are shown. The sound speed in water is 1440 m/s and the RMS roughness is 0.01 m. It is seen that a rough bottom acts effectively as a low-pass filter on the reflected signals.

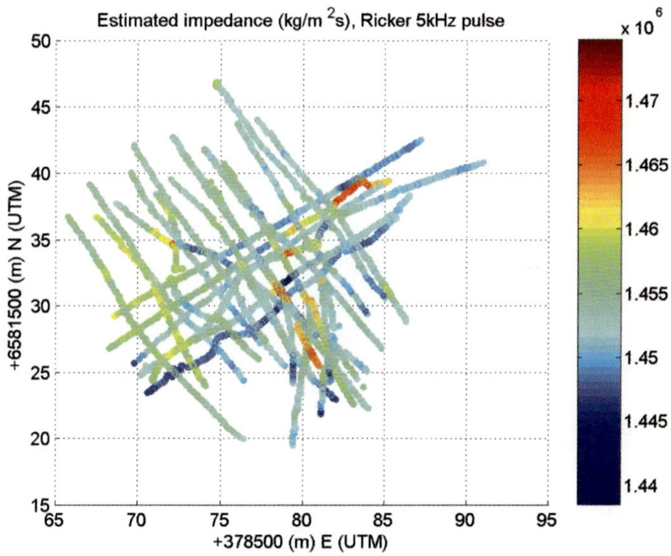

Figure 12.6. FARIM analysis: estimated impedance not corrected for roughness, 5-kHz pulse.

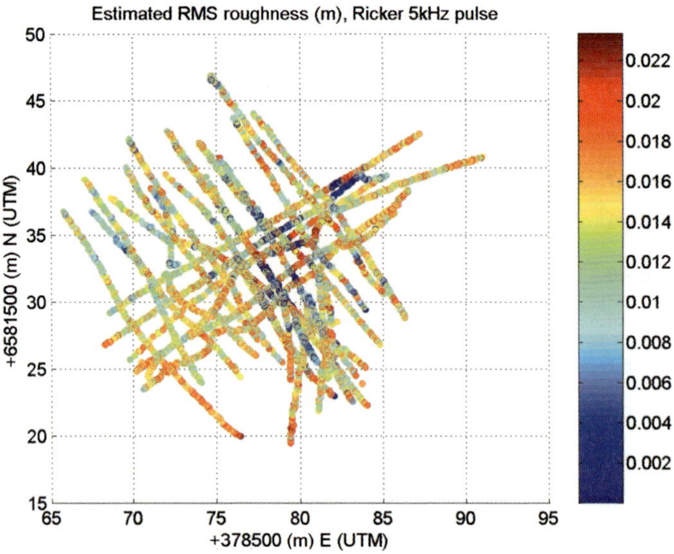

Figure 12.7. FARIM analysis: estimated RMS roughness, 5-kHz pulse.

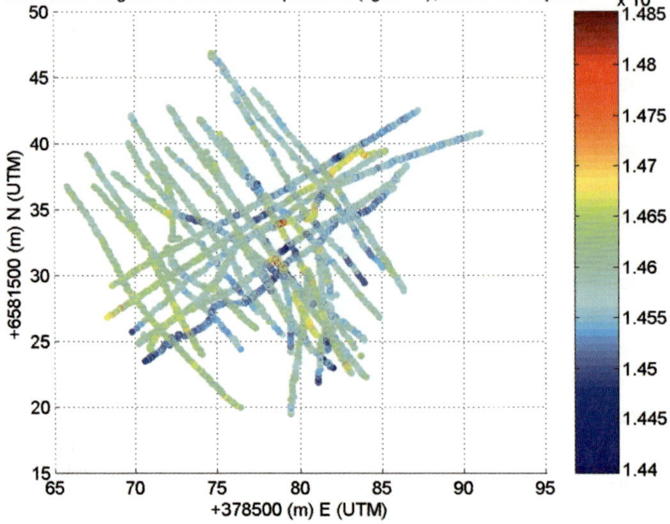

Figure 12.8. FARIM analysis: estimated impedance corrected for roughness, 5-kHz pulse.

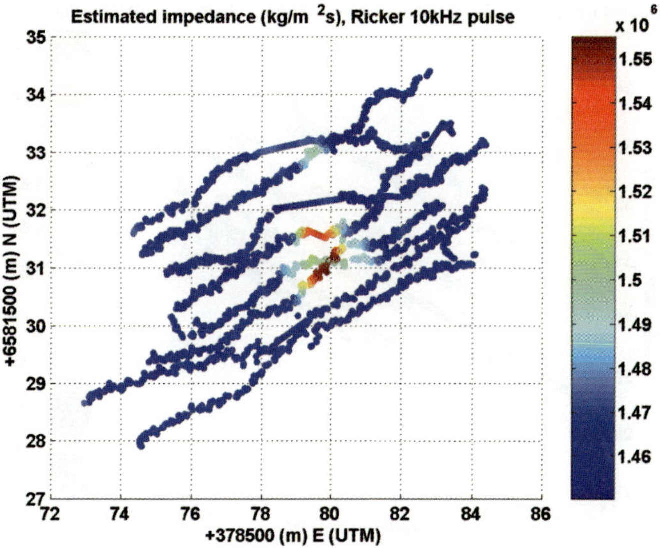

Figure 12.9. FARIM analysis: estimated impedance not corrected for roughness, 10-kHz pulse.

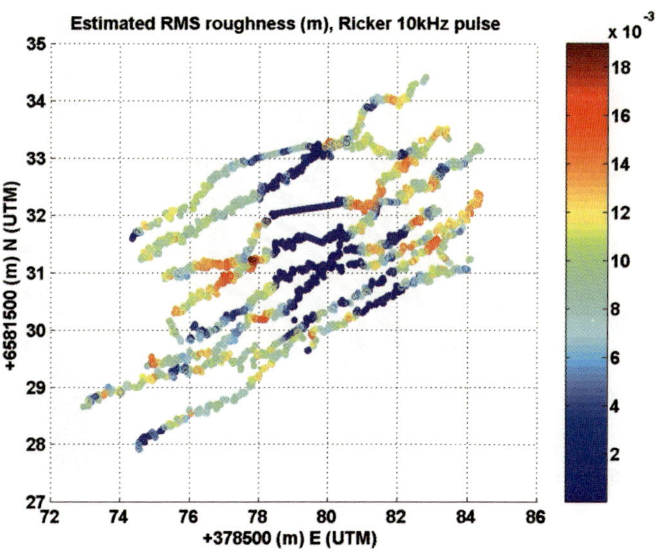

Figure 12.10. FARIM analysis: estimated RMS roughness, 10-kHz pulse.

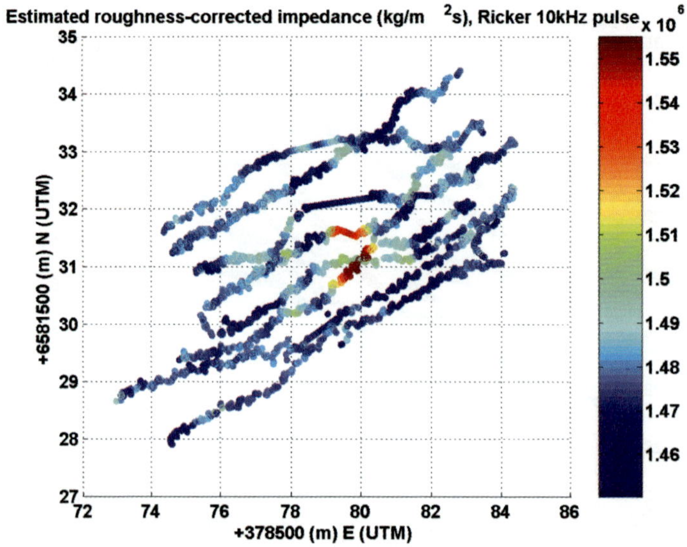

Figure 12.11. FARIM analysis: estimated impedance corrected for roughness, 10-kHz pulse.

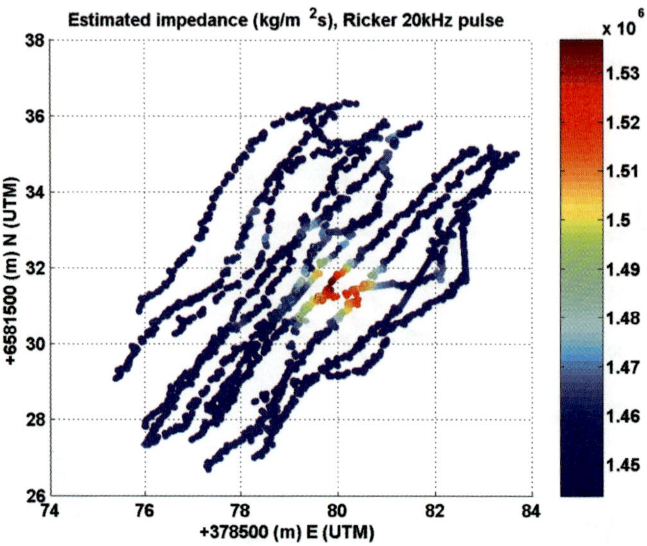

Figure 12.12. FARIM analysis: estimated impedance not corrected for roughness, 20-kHz pulse.

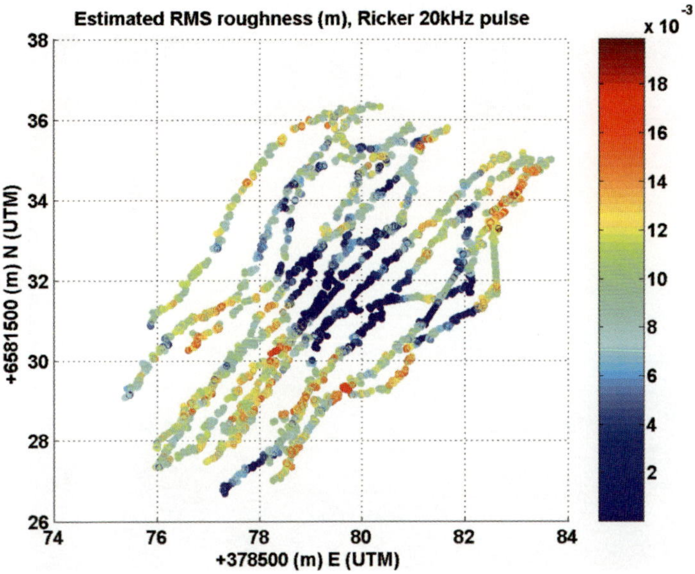

Figure 12.13. FARIM analysis: estimated RMS roughness, 20-kHz pulse.

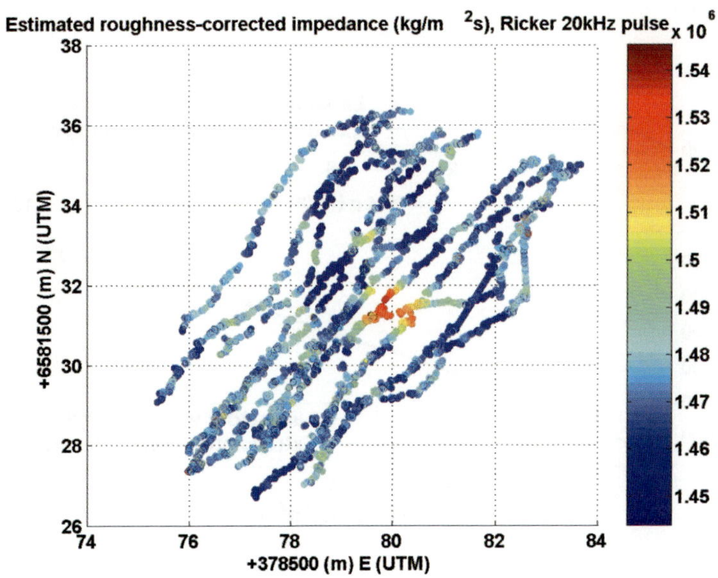

Figure 12.14. FARIM analysis: estimated impedance corrected for roughness, 20-kHz pulse.

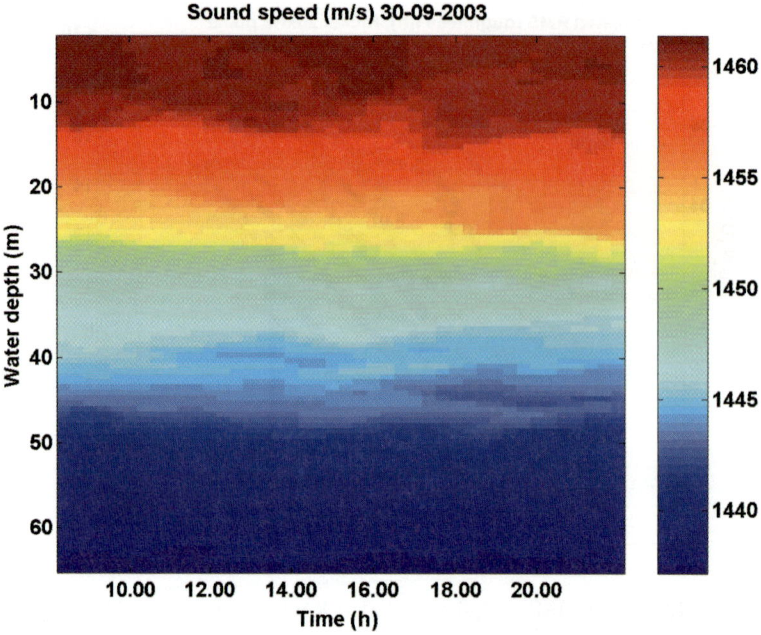

Figure 12.15. Measured sound speed in water during one of the days in the sea trial.

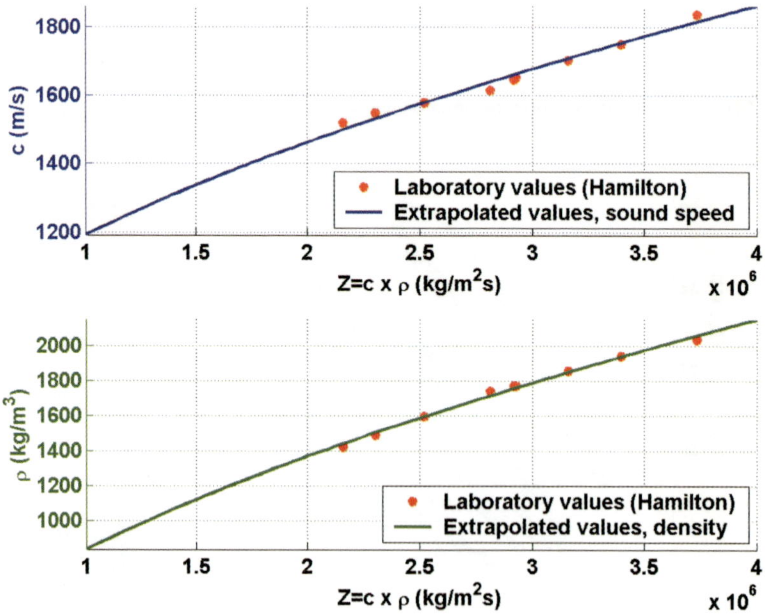

Figure 12.16. Linear regression between sound speed and density based on measurements in laboratory for different sediment types. The sound speed and density can thus be determined from an estimate of sea bottom impedance.

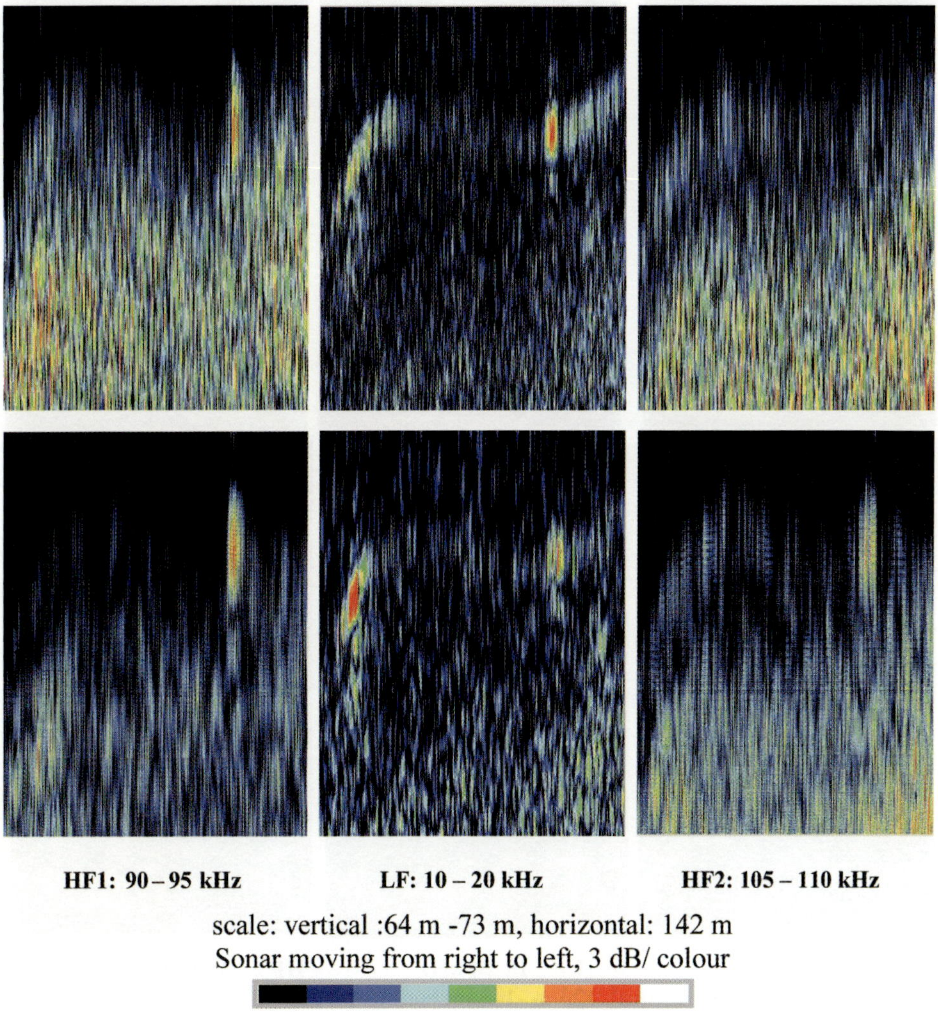

Figure 13.1. Isolated bright-spot area: raw data (top) and synthetic aperture processed data (bottom).

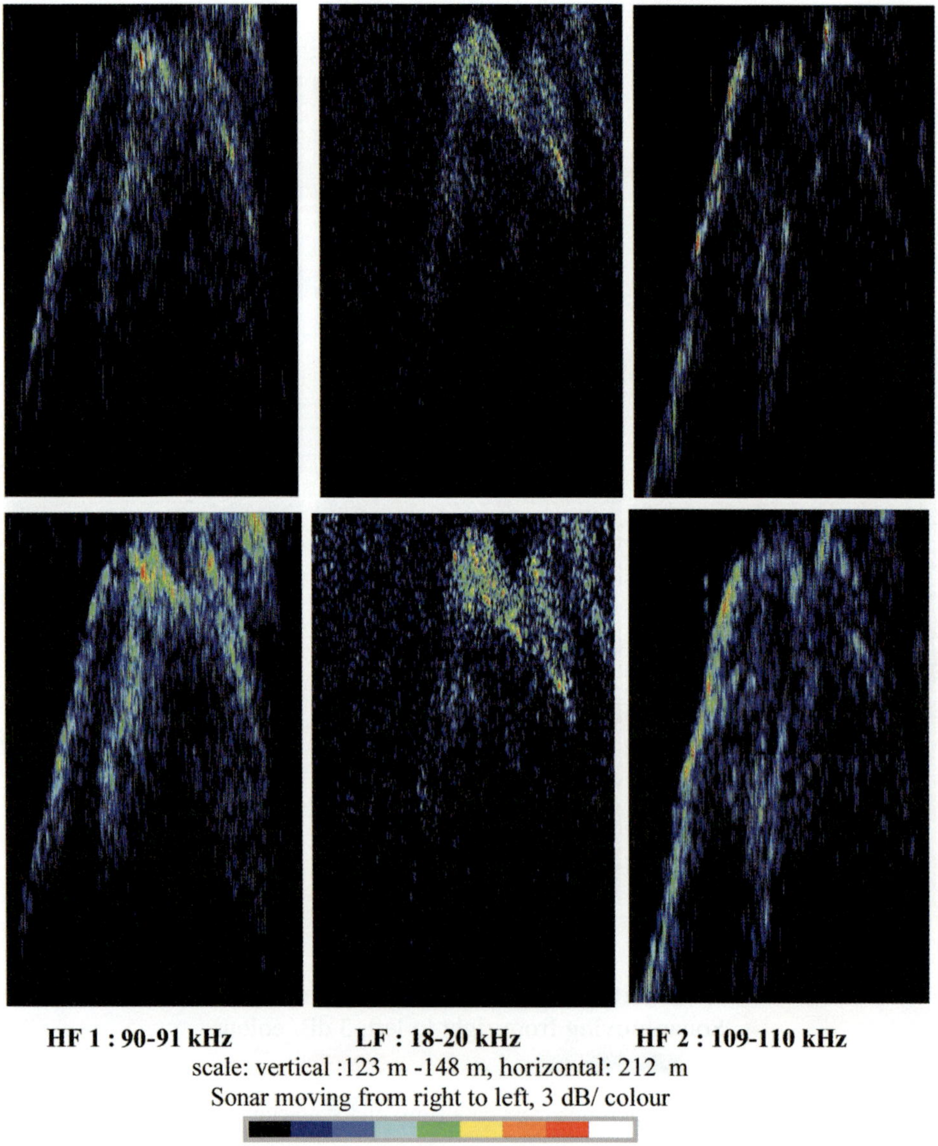

Figure 13.2. "Reverberating" area: raw data (top) and synthetic aperature processed data (bottom).

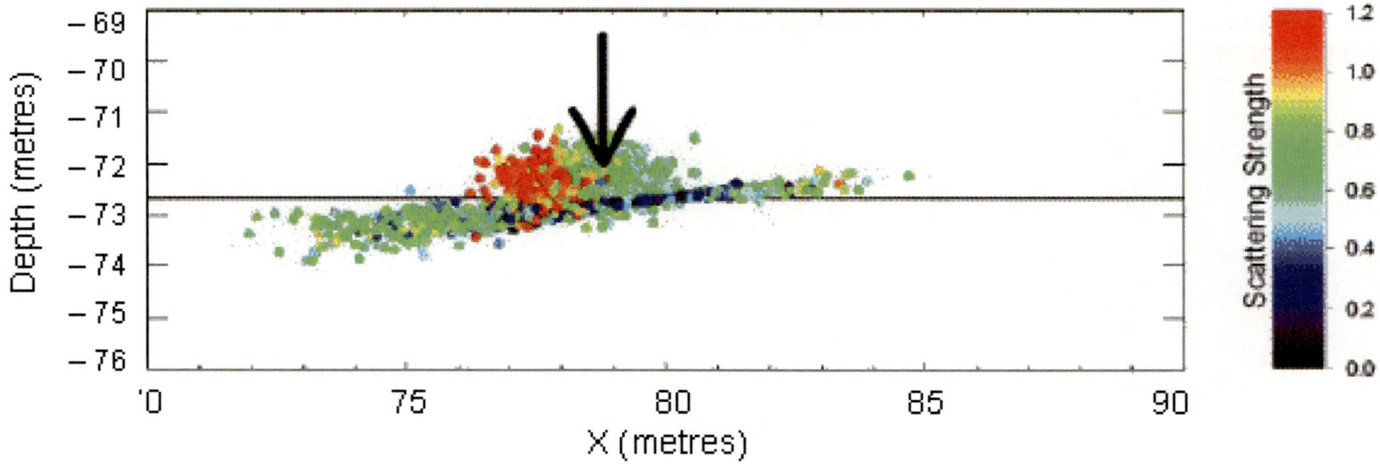

Figure 14.4. Scattering points as determined from the localization routine, projected in the x-depth plane: the colours represent relative scattering strength, the arrow indicates the nominal box location and the black line indicates the nominal seabed depth.

(a)

(b)

(c)

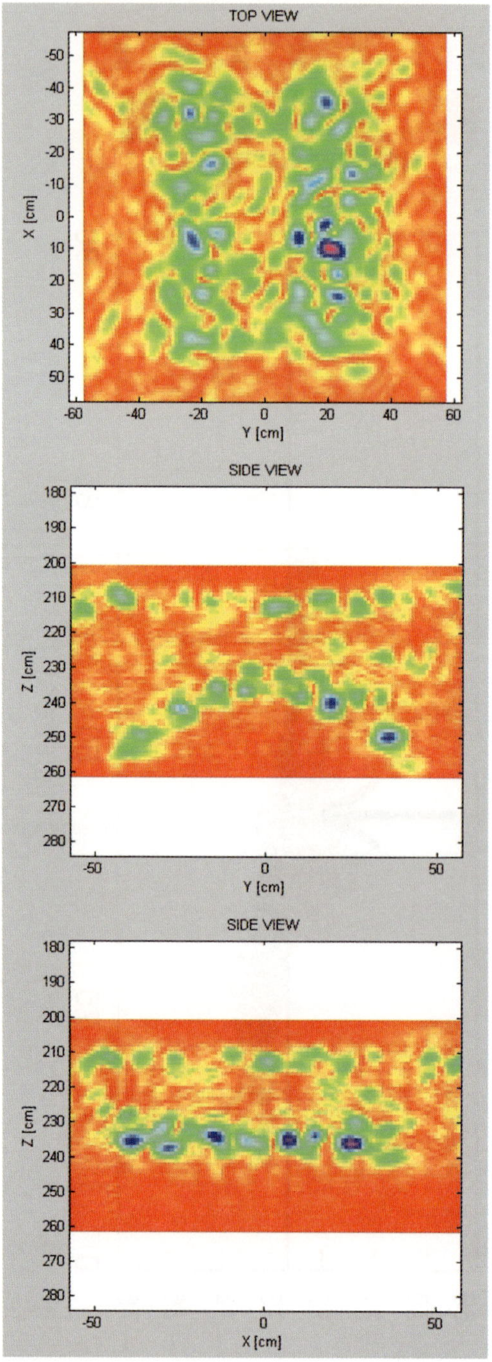

Figure 16.2. Simulated data with linear visualization: (a) slice parallel to plane $z = 0$; (b) slice parallel to plane $x = 0$; and (c) slice parallel to plane $y = 0$.

(a)

(b)

(c)

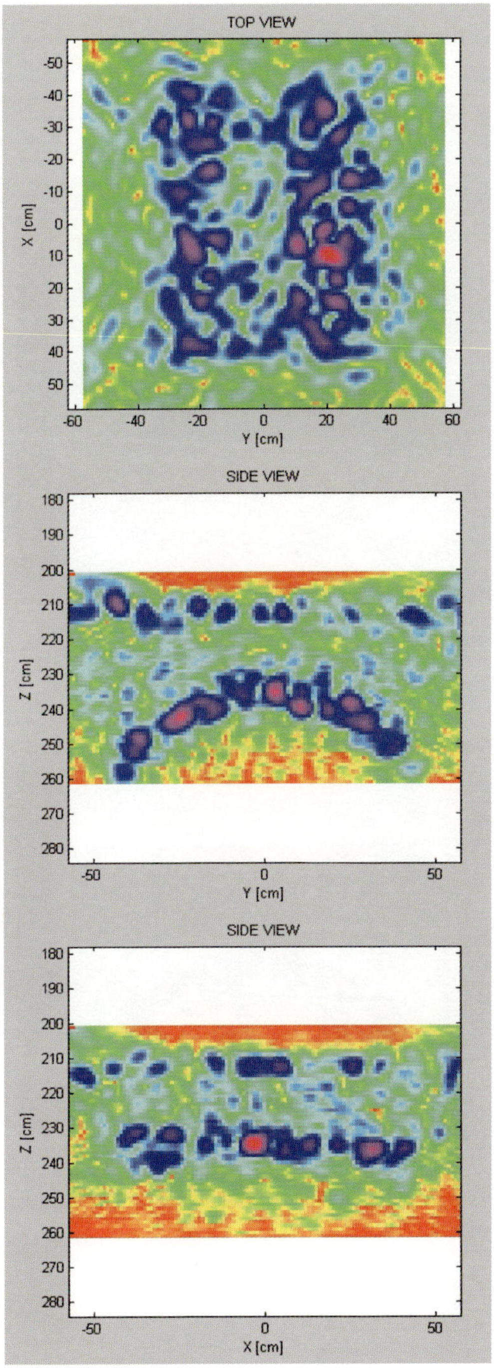

Figure 16.3. Simulated data with dB visualization: (a) slice parallel to plane $z = 0$; (b) slice parallel to plane $x = 0$; and (c) slice parallel to plane $y = 0$.

(a)

(b)

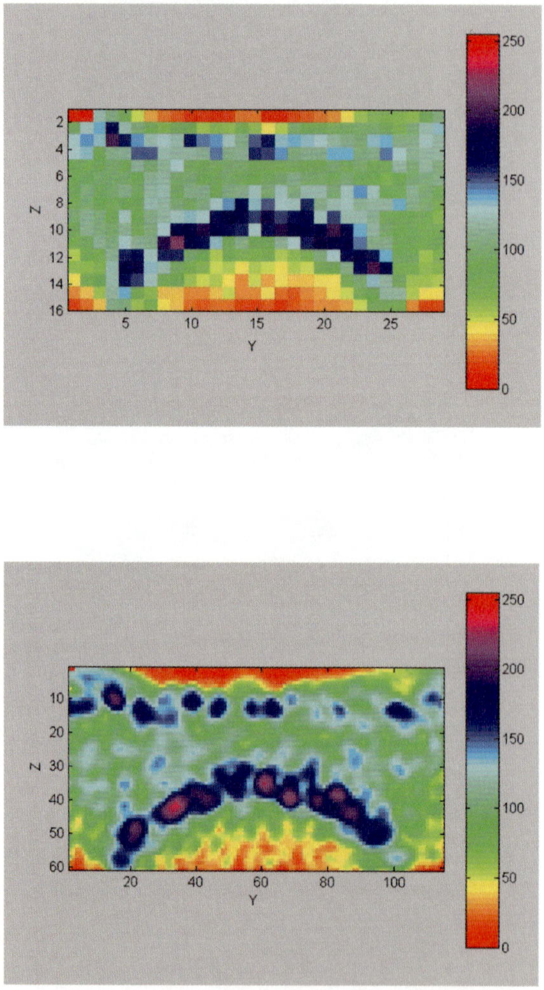

Figure 16.4. Pre-processing results: (a) slice of the octree level-2 image; (b) slice of the 3D image filtered by a $3 \times 3 \times 3$ median filter.

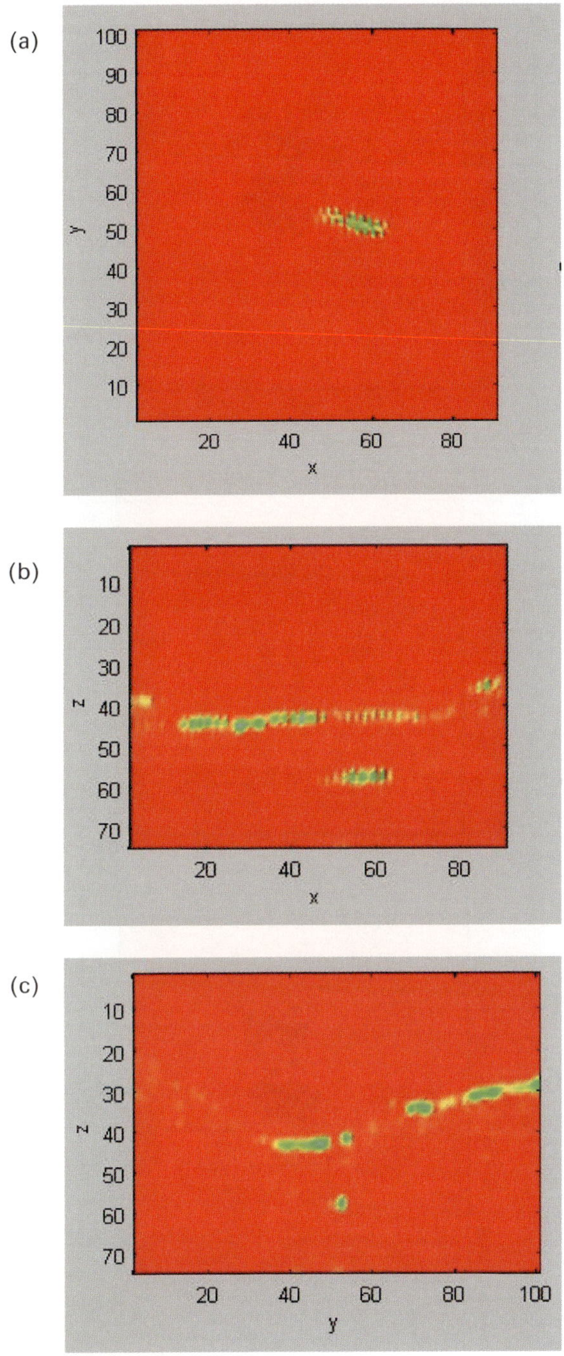

Figure 16.15. Real raw data visualization: (a) slice parallel to plane $z = 0$; (b) slice parallel to plane $y = 0$; and (c) slice parallel to plane $x = 0$.

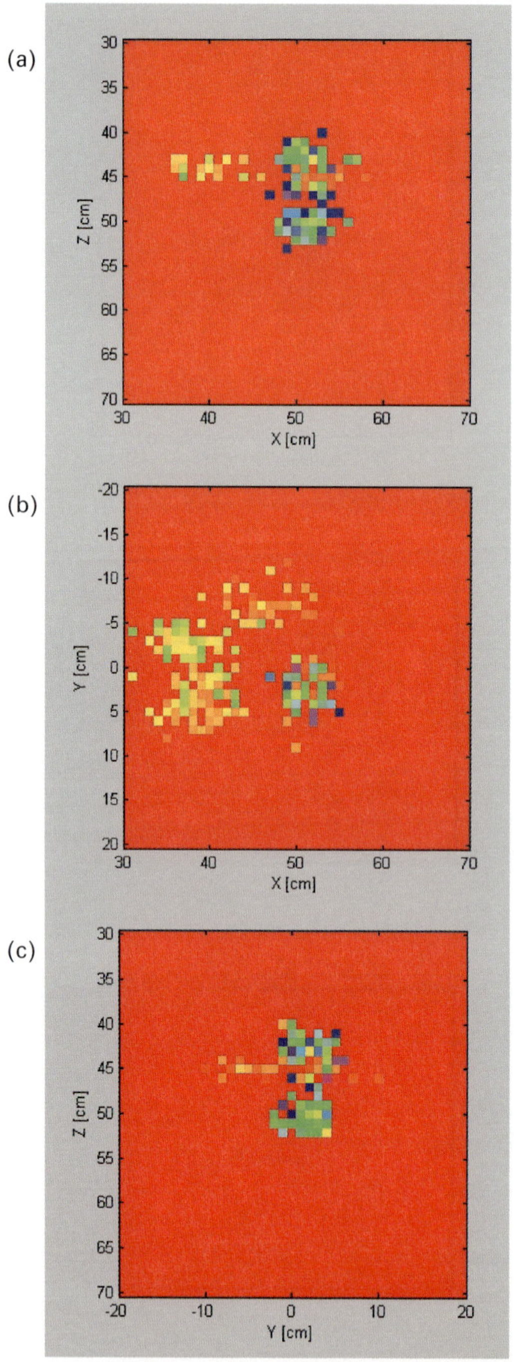

Figure 16.19. Raw data visualization: (a) slice parallel to plane $y = 0$; (b) slice parallel to plane $z = 0$; and (c) slice parallel to plane $x = 0$.

(a)

(b)

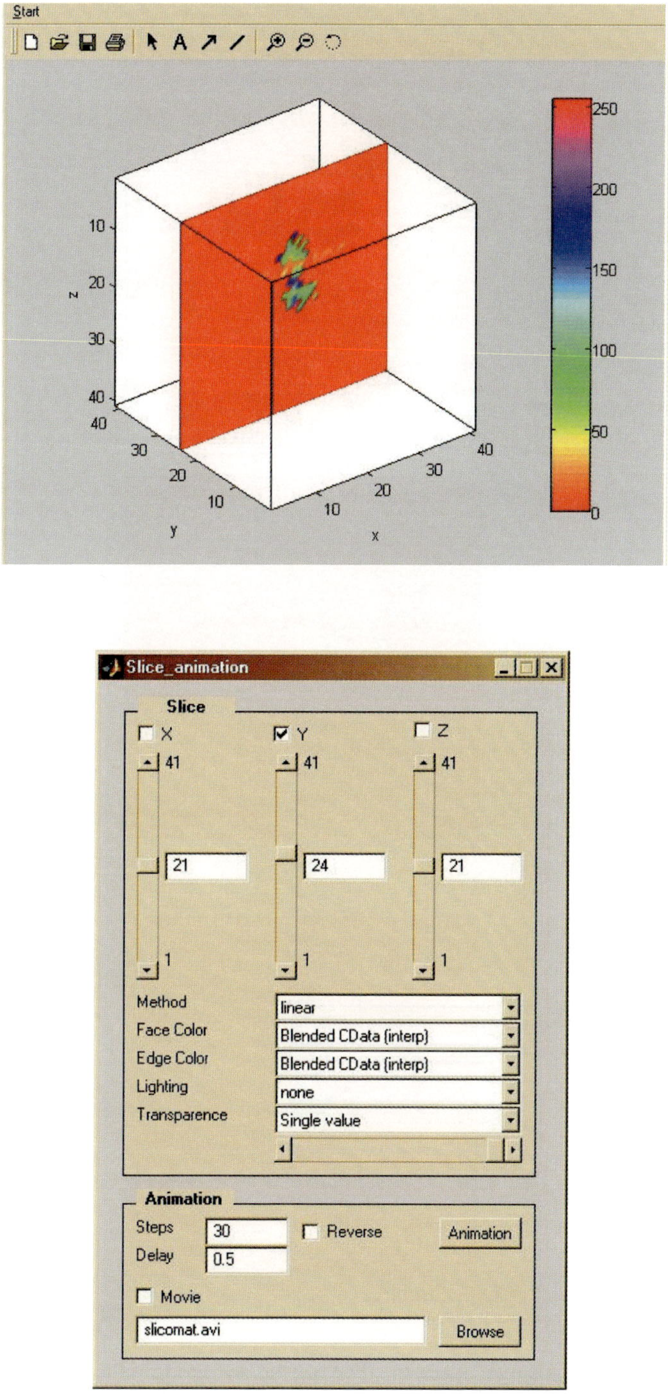

Figure 16.20. (a) 3D visualization of the raw data; and (b) toolbox menu.

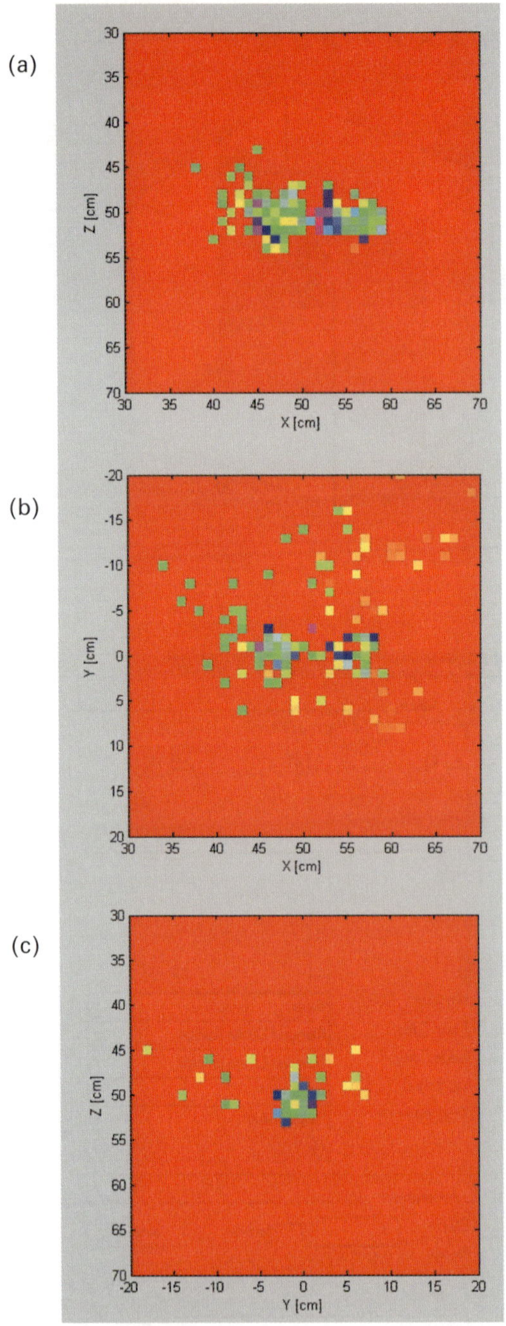

Figure 16.25. Two-dimensional slices of 3D acoustic scattering data from the MAS experimental technique: (a) slice parallel to plane $y = 0$; (b) slice parallel to plane $z = 0$; and (c) slice parallel to plane $x = 0$.

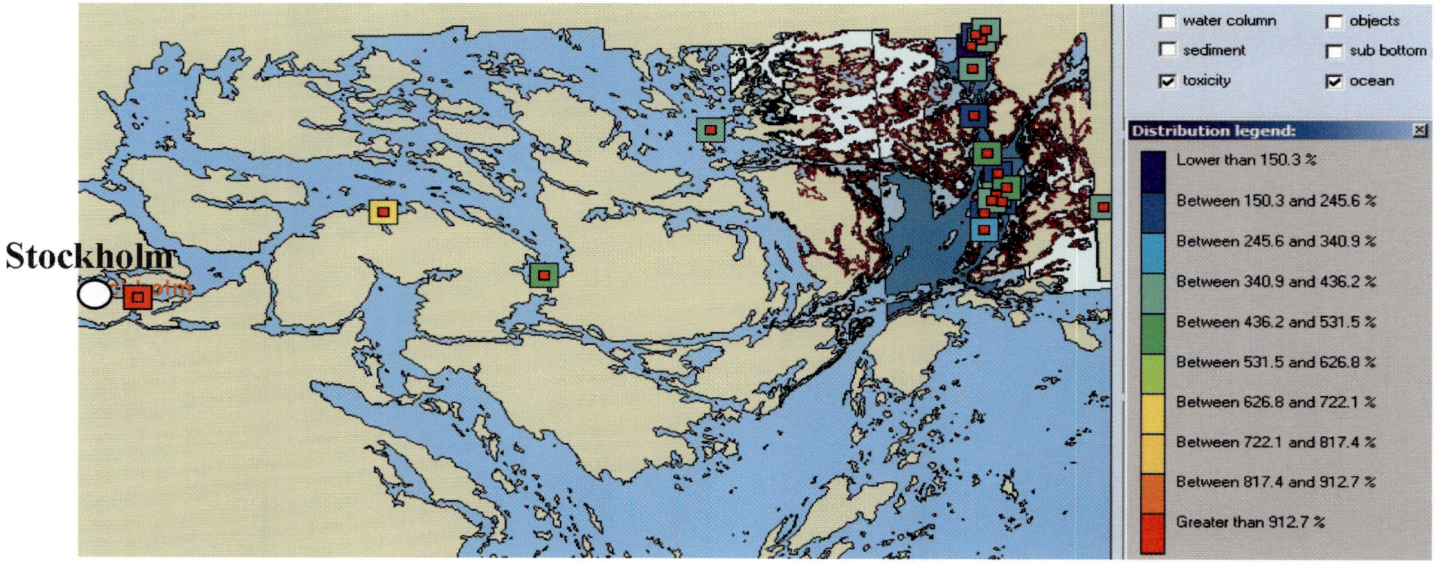

Figure 17.1. Thematic map of EROD-induced activity built from the data reported in Figure 15.12, as displayed on the SITAR GIS interface; each square corresponds to a sample point, and the colour-coding detailed in the scale legend goes from red (most environmentally harmful levels) to dark blue (less environmentally harmful levels). The SITAR investigation area in Möja Söderfjärd is at the right side of the map, where sample density is higher.

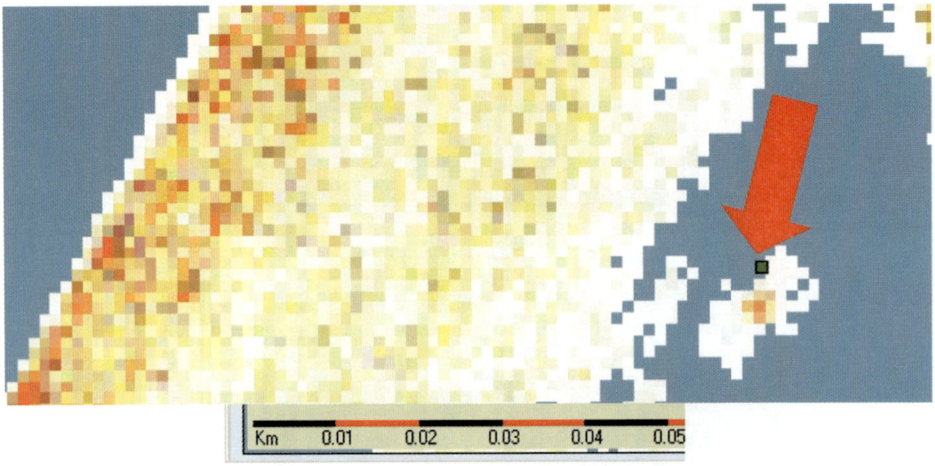

Figure 19.1. PSSS high-frequency (110-kHz) run: ground truth data indicate that an object (dark green square, indicated by the red arrow) is present in correspondence to an isolated feature in the PSSS data. The scale at the bottom of the figure corresponds to a 50-m length.

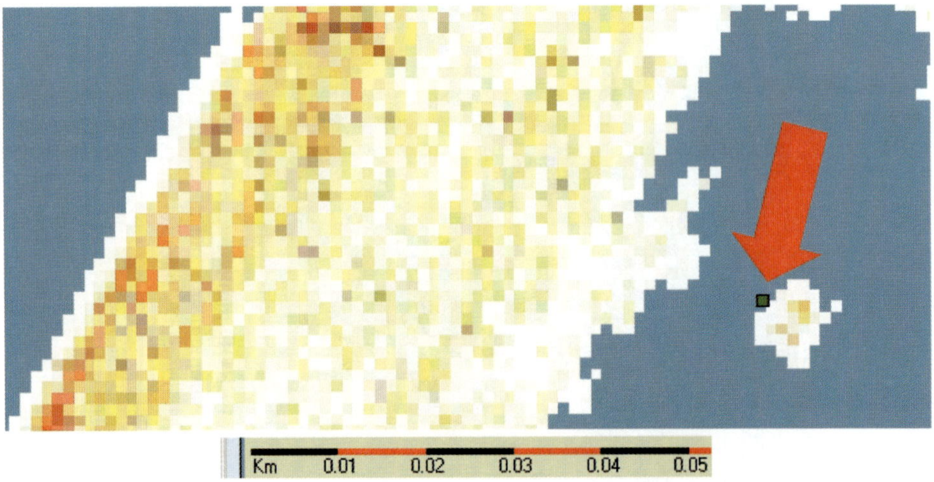

Figure 19.2. Data from Figure 19.1 after synthetic aperture processing. The isolated feature of Figure 19.1 appears now as two distinct features. Historical data ("ground truth") reported – in approximately the same position – a single object (dark green square, indicated by the red arrow). The scale at the bottom of the figure corresponds to a 50-m length.

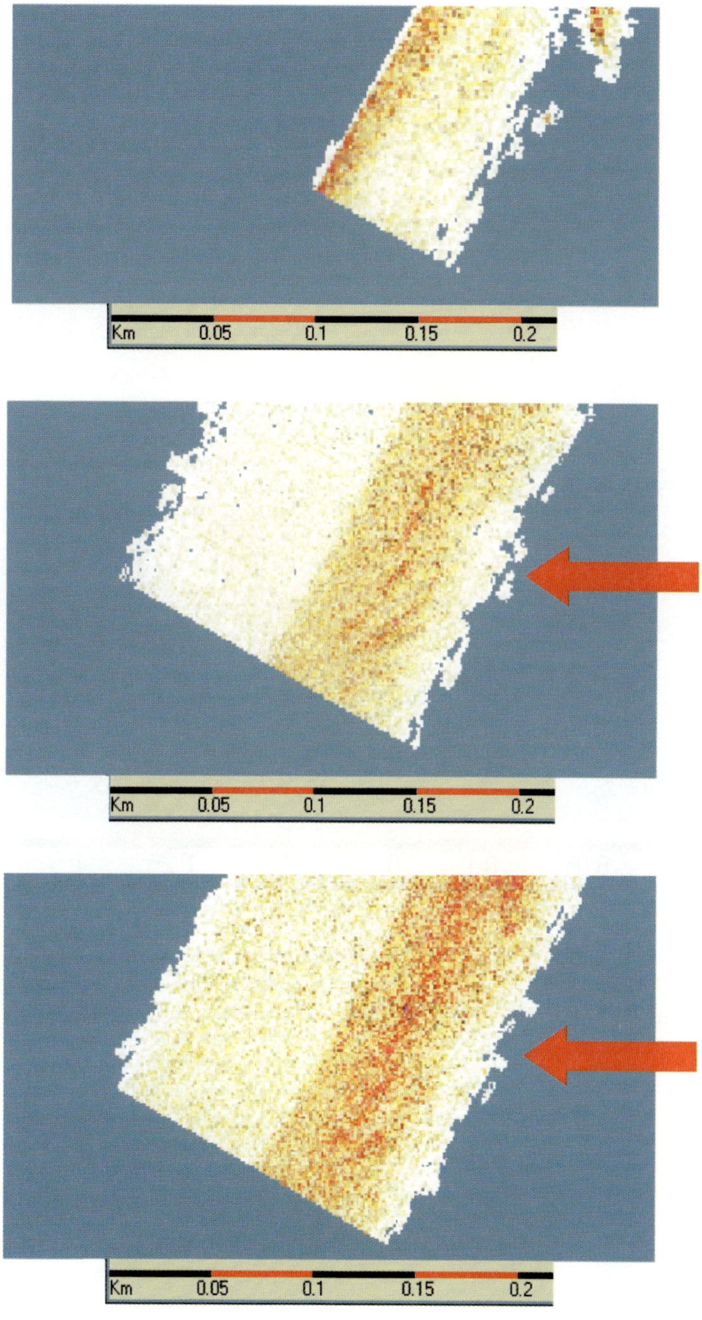

Figure 19.3. From top to bottom, high-frequency (110-kHz) PSSS data, low-frequency (20-kHz) data and low-frequency (20-kHz) data after synthetic aperture processing of the same portion of seafloor. The scale at the bottom of each figure is 200 m long. The low-frequency reverberating feature is indicated by the red arrow.

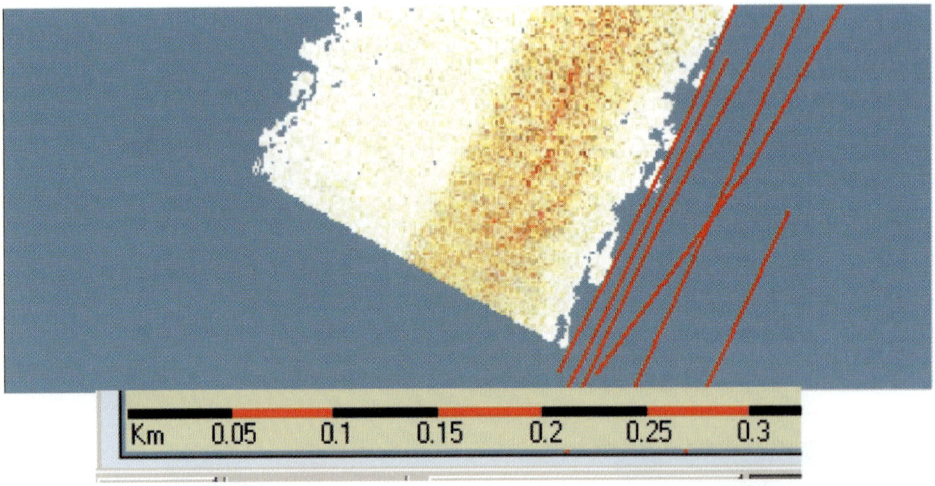

Figure 19.4. PSSS data (low frequency) with reverberating features and position of seismic lines ("ground truth" – historical data) available from the Swedish Geological Survey. Closest line to the reverberating feature is about 50 m distant.

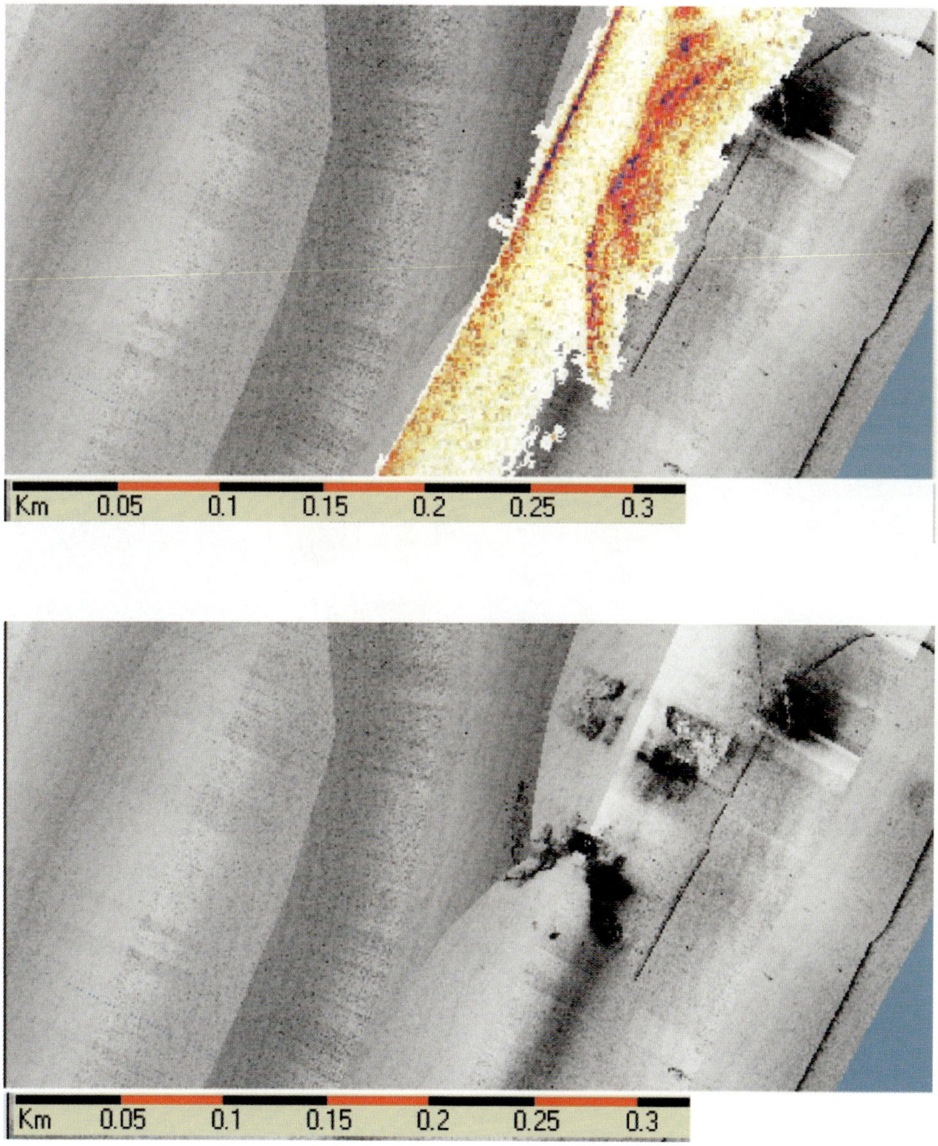

Figure 19.6. Comparison between PSSS high-frequency (110-kHz) data (top, hot colours) and Klein standard side-scan sonar data (bottom, grey-level shading). Klein data are the result of mosaicking different transects. Scale at the bottom of each figure has a 300-m length.

Figure 19.7. Comparison between PSSS high-frequency (110-kHz) data and bathymetry. PSSS data are the same data as in Figure 19.6. The PSSS image does match the very fine bathymetric details of the area.

13

Buried waste localization: the parametric, synthetic aperture, side-scan sonar

M. Zakharia and J. Dybedal

13.1 INTRODUCTION

The results obtained using the parametric side-scan sonar (PSSS) instrument are shown in this chapter with respect to two kinds of acoustic features that appear in various places in the field dataset: these features are here termed "isolated bright spots" and "reverberating areas". From an interpretation point of view, "bright spots" may be associated with isolated, small-scale, geological features or to man-made objects, while reverberating areas are very likely associated with geological features. Some additional PSSS data, substantiating this interpretation, are shown in Part V (Evaluation). Only those lines in the dataset that show – by visual inspection – very prominent acoustic features, after initial pre-processing, have been analysed so far; the dataset is very large, and data analysis is likely to proceed well after the conclusion of the project. It has finally to be remarked that the PSSS was operating in the field test in a sub-optimal configuration, due to wiring problems that were brought into evidence only in the subsequent full-scale tank test data. In this respect, the results reported here could well be improved with an updated instrument configuration.

13.2 ISOLATED BRIGHT SPOTS

In order to easily demonstrate the gain in resolution, image examples from sea trials, where bright spots were present, are first shown in Figure 13.1 (see colour section). The transmitted signals used in the presented data are linear FM sweeps or chirps:[1] 95–90 kHz, 10–20 kHz and 105–110 kHz from left to right, respectively.

The "bright spot", which probably is a target located on the sea floor, is seen only in the 90–95-kHz and 10–20-kHz raw data. However, it appears also in the

[1] A chirp is a time-limited sinusoidal signal.

105–110-kHz data after synthetic aperture processing; see the lower, right-hand part of Figure 13.1. Note that another "bright spot" appears in the low-frequency data after synthetic aperture processing. This echo came from a buried target, which, unfortunately, was not verified. The high frequency signal did not penetrate sufficiently to give a good echo from the same target. Discrimination between the buried target and proud ones was confirmed by interferometric processing.

13.3 REVERBERATING AREA

The same processing schemes were applied to data from the reverberating area (Figure 13.2, see colour section). Although the three images are obtained simultaneously, several differences between the frequency ranges can be observed. Comparing HF and LF images illustrates the difference in the echo formation mechanism in both frequency ranges. In the high-frequency cases, the speckle due to interface scattering is very high. The speckle is less visible in the LF image and other features are visible that may correspond to sub-bottom volume scattering. A close look at the high-frequency images shows that, although they look very similar, several differences exist between both images. The difference in level may be due to a probable asymmetry of the transfer function of transmit–receive arrays around 100 kHz. The only explanation we could find to explain feature differences (high-space and -frequency variability in surface backscattering strength) is that we were dealing with a gassy sediment where bubble resonance could lead to high-frequency variability. It is worth noting that the central image (LF) possesses a better axial resolution than those at the side, as the LF modulation bandwidth is twice that of the HF.

Preliminary analysis of these images obtained using a parametric synthetic aperture sonar showed both the relevance of the methodology and the complexity (and the richness) of wideband echo formation mechanisms. First results showed large differences between LF and HF images and even between HF images themselves, although their central frequencies were close.

14

Buried waste inspection: acoustical images and inversion from multiple-aspect scattering

I. Karasalo, P. Skogqvist, Ph. Blondel and P. F. Dobbins

14.1 INTRODUCTION

The multiple-aspect scattering (MAS) dataset collected at the field test site in Möja Söderfjärd concentrated on the half-buried box in a known location, as described in Section 11.3. As reported there, due to equipment malfunction, the data have been gathered using the spare vertical array in a configuration different from the one originally planned (fewer receivers, different spacing and elevation above the sub-bottom). From the point of view of 3D acoustical imaging, this had a negative effect: the scatterers identified from the object in this configuration were not dense enough, and had too much uncertainty, to produce a volumetric 3D acoustic image to be used for extraction of object information through decision support system processing. On the other hand, the identified scatterers were sufficient to produce a 2D image in which the object itself and its position could be determined with remarkable accuracy. Moreover, the raw acoustic data were used, in conjunction with the acoustic scattering models described in Chapter 4, to perform a model-based inversion for determination of the scatterer geometrical parameters. The success of this latter approach demonstrates that, even in sub-optimal configurations, the MAS technique does provide sufficient information for the acoustical inspection of buried and proud objects.

In the following we concentrate briefly on 3D acoustical image processing and then, in more detail, on estimation of object parameters.

14.2 3D ACOUSTICAL IMAGES

Multiple-aspect scattering measurements of the half-buried box were processed using the procedures outlined in Section 8.3 (*Target identification and reconstruction*). Particular care had to be given to the exact determination of the position and attitude of the ROV with the imaging sonar, as well as the exact synchronization of transmission

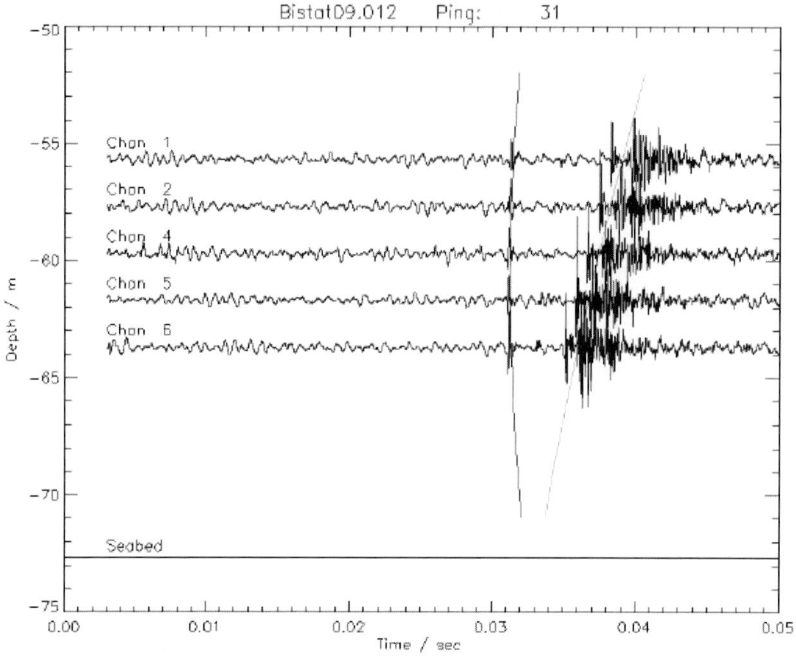

Figure 14.1. Signals at five hydrophones plotted at a height representing the hydrophone depth. The curve on the left shows the estimated direct path arrival time and the curve on the right is the estimated surface reflection, both plotted against receiver depth (Dobbins and Blondel, 2004).

with reception (this sometimes proved to be difficult to assess in real time). Figure 14.1 shows one example from the many scattered acoustic signals received as a function of time, and the typical variations according to the receiving hydrophone (and therefore the scattering angle).

Deconvolution through adaptive Wiener filtering was applied to both Ricker[1] and chirp[2] pulses transmitted in bistatic mode. Figures 14.2 and 14.3 report, respectively, the received signals and the results of the deconvolution process and target localization procedure.

An example of localization of scattering points is given in Figure 14.4 (see colour section); in this figure, which is a projection on the x-depth plane of the 3D data, the box-shaped target penetrating the seabed can be recognized. However, the uncertainties in the localization process, due to the availability of a reduced set of receivers with respect to what was planned, did not allow for the reconstruction of a full 3D acoustic image, similar to those obtained from the tank experiment reported in Section 16.2.

[1] A Ricker signal is a specific, pulse-like, signal shape very often used in marine seismic exploration.
[2] A chirp is a time-limited sinusoidal signal.

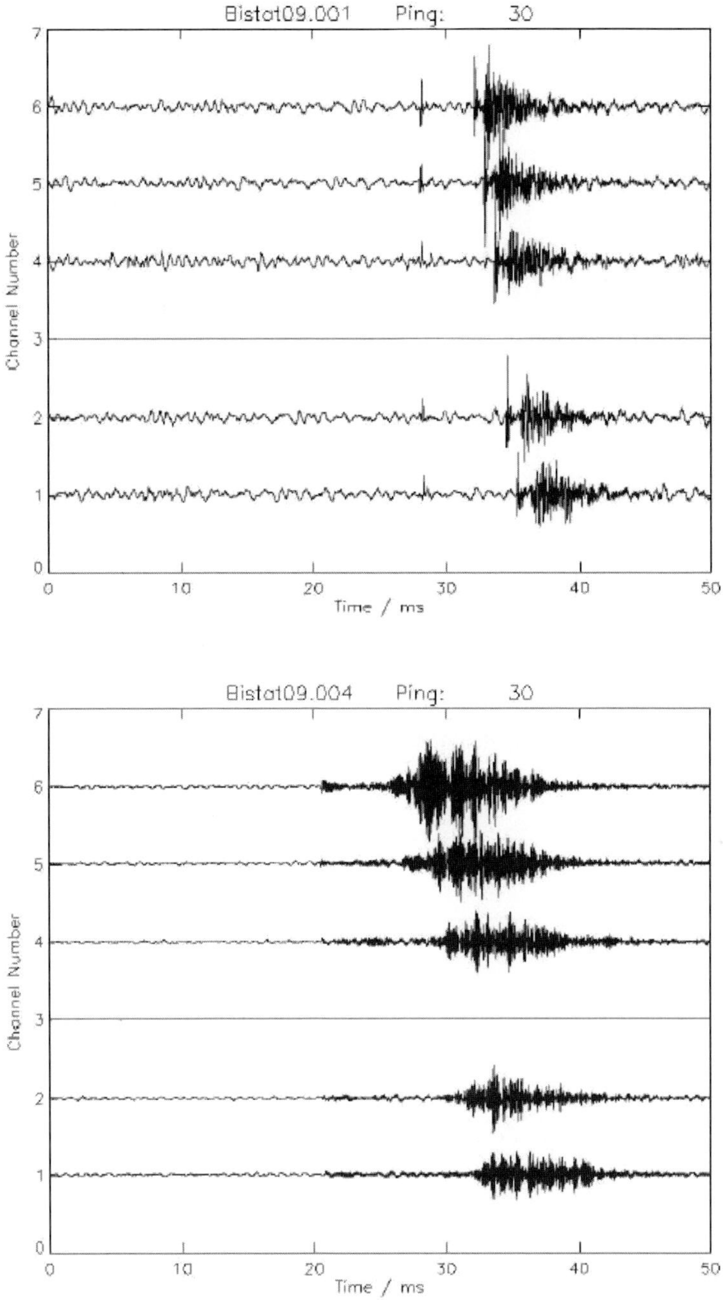

Figure 14.2. Raw data from MAS bistatic survey acquired through transmission of a 10-kHz Ricker pulse (top), and of a 10–20-kHz chirp signal of 2.5-ms duration (bottom); receiver gains were set at a lower level in chirp mode, and channel 3 was not operational at the time of acquisition.

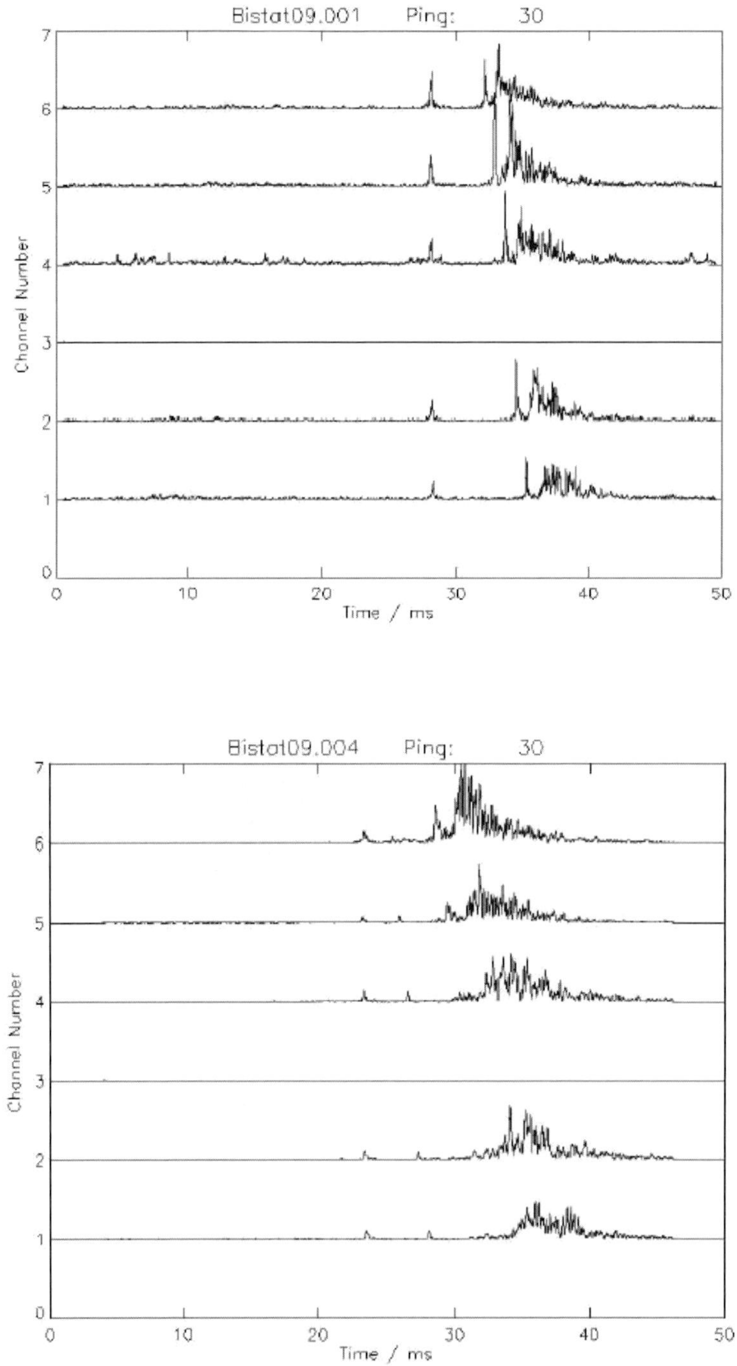

Figure 14.3. Results of the deconvolution process of the data of Figure 14.2: from Ricker-generated data (top), and from chirp-generated data (bottom).

14.3 ESTIMATION OF OBJECT PARAMETERS

We consider the problem of estimating a set of parameters for a scattering object, such as size, shape, orientation, sound speed and density, from measurements of the scattered field. We denote by **u** a vector of such parameters in a model of the scattering scenario, and by $p(\mathbf{u}) = p(x, t, u)$ the corresponding model-predicted scattered field, computable by the methods described above. Denoting the experimentally observed field by $q = q(x, t)$, the identification problem is then to find u such that a fitness function $\Phi(\mathbf{u})$ measuring the distance between $p(\mathbf{u})$ and q is minimized.

The top frame of Figure 14.5 shows a section of experimentally observed signals at receivers 2 to 6. The section contains (i) the specular echo from the upper surface of the box-shaped scatterer, followed by (ii) a ca. 1-ms-long low-amplitude period of returns from a very soft top sediment layer and (iii) stronger returns from underlying seabed layers. The bottom frame shows simulated echoes computed using the Ray–Kirchhoff (RK – see Part I) method (solid) together with the specular echo intervals of the experimental traces (dashed). The dashed cross-trace curves show the arrival times of the echo reflected at the upper surface of the box, as predicted by ray theory.

The arrival times of model-predicted echoes were seen to agree well with the experimental data. However, their amplitudes as a function of receiver depth were nearly uniform, with a weakly pronounced maximum at receiver 4, the depth at which specular reflection occurs at the flat upper surface of the box. In contrast, the experimentally observed echo amplitude increased with depth and was maximal at the deepest receiver.

Motivated by the characteristics of the experimental data and the accuracy properties of the RK-predicted transient echoes the fitness function $\Phi(\mathbf{u})$ for the parameter inversion was defined as

$$\Phi(\mathbf{u}) = \sum_{i=1}^{N_{rec}} |T_i(\mathbf{u}) - \tau_i|^2 + |A_i(\mathbf{u}) - \alpha_i|^2, \tag{14.1}$$

where $T_i(\mathbf{u})$ and $A_i(\mathbf{u})$ are the computed arrival time and the normalized amplitude at receiver i; τ_i and α_i are the measured arrival time and the normalized measured amplitude at receiver i.

Figure 14.6 shows two examples of the fitness function (14.1) as a function of two parameters of the box-shaped scatterer: the roll and the yaw angles (left) and the x- and z-coordinates of the centre point (right).

In general $\Phi(u)$ is a complicated nonlinear function, with multiple local minima, and the optimal parameter vector $\hat{u}$ must be sought from methods for global minimization. In this study two such algorithms were considered: a differential evolution algorithm (DE – Storn and Price, 1995) and a hierarchical genetic algorithm (GA – Cantú-Pas, 1998), both combined with a final search of local minima by the downhill simplex method (Nelder and Mead, 1965). The RK method described was used as the forward model for computing the scattered field $p(u) = p(x, t, u)$.

As a first step towards estimating the parameters of the scatterer, the methods described above were applied to calculate the roll and yaw angles of rotation of the

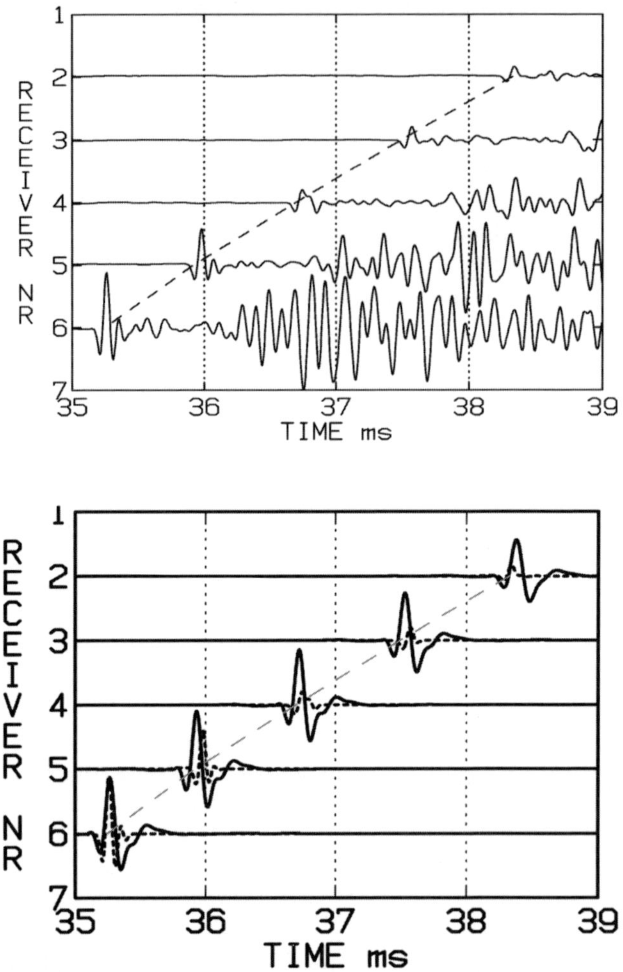

Figure 14.5. Signals in the middle five hydrophones of the vertical array: (top) experimental data; (bottom) predicted by the RK method (solid) and experimental data (dashed). Scatterer parameters based on prior information.

scatterer, keeping other parameters fixed. Table 14.1 shows the results of the minimization, the number of iterations and the number of function evaluations needed to locate a local minimum.

The GA method actually finds the minimum after only 7 iterations in this run, but the standard deviation in the population is in that stage still too high to terminate the search. The DE method needs 17 iterations to find the minimum and is terminated after 27 iterations by the condition that the population has not changed in 10 consecutive iterations. Figure 14.7 shows the fitness function $\Phi(\mathbf{u})$, with the best individual in each generation marked by dots. $\Phi(\mathbf{u})$ has two local minima in the

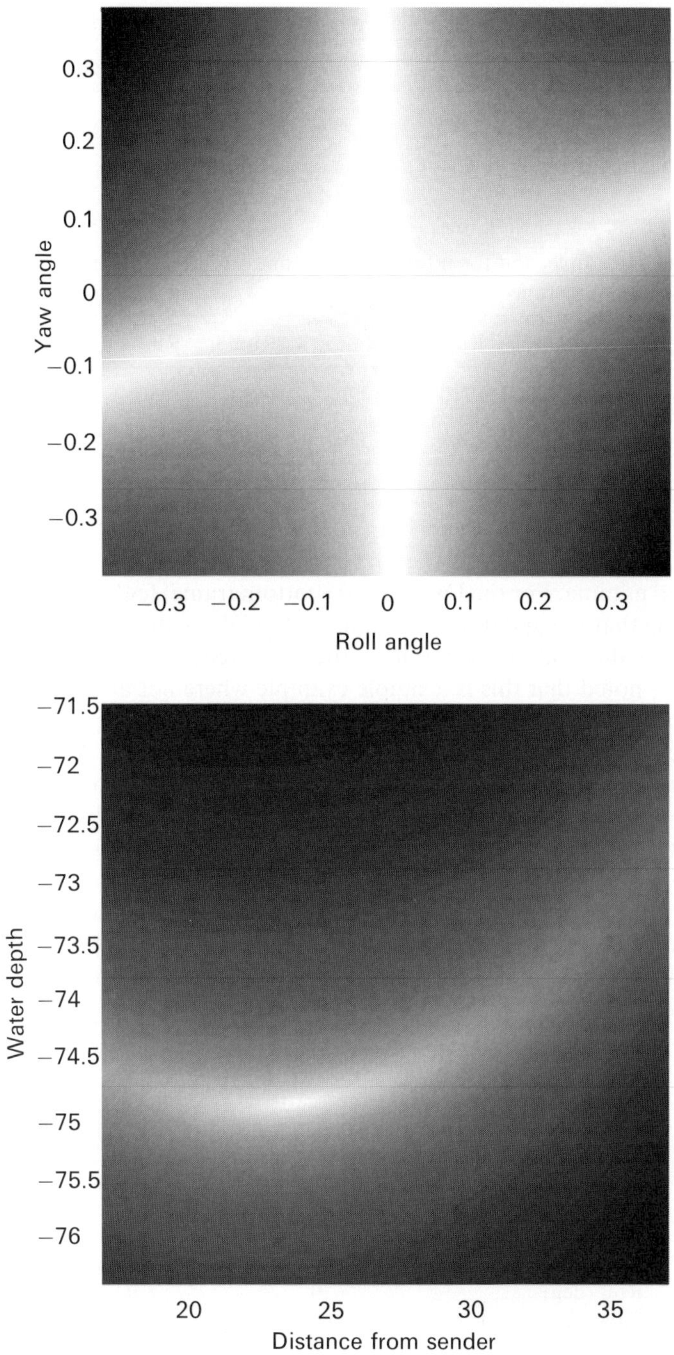

Figure 14.6. Fitness function Φ in Equation (14.1): (top) as a function of roll and yaw angle; (bottom) as a function of the x- and z-coordinates of the centre point.

Table 14.1. Results of global search for roll and yaw angles.

	Roll	Yaw	Iterations	Function evaluations
GA	−0.86	11.8	40	800
DE	−0.47	10.1	25	450

Table 14.2. Inversion for interior density and sound speed. Inversion results are underlined.

	Run 1	Run 2	Run 3	Run 4	Run 5
Density (kg/m^3)	<u>1740</u>	1420	1490	1630	<u>1490</u>
Sound speed (m/s)	2680	2570	<u>2530</u>	<u>2380</u>	<u>2570</u>

region shown. The individuals of the GA method (top frame) are seen to cluster around the local minima. For the DE method (bottom frame) fewer than 25 values are shown, reflecting that the updating procedure of the DE method leaves an individual unchanged unless the trial individual has a better fitness.

It should be noted that this is a simple example where not all the benefits of the global search methods come into play; a sound local minimizer will probably converge faster with the global minimum.

As a second step, the interior density and sound velocity of the box were estimated in five similar runs, with results as shown in Table 14.2. The underlined values are those resulting from the parameter search, other entries show values kept fixed during the search.

Similarly, in a third and a fourth step, the depth and range of the centre of the box and the pitch angle of the box, respectively, were inverted, keeping other parameters fixed at their values from previous inversions. The results of all inversion runs are collected in Table 14.3.

Table 14.3. Parameters of the scatterer determined by inversion.

	Initial	Inverted
Range (m)	24	23.95
Depth (m)	74.85	74.85
Roll (deg)	0	0.9
Pitch (deg)	0	4.8
Yaw (deg)	19	11.7
Density (kg/m^3)	1630	1490
Sound speed (m/s)	2680	2570

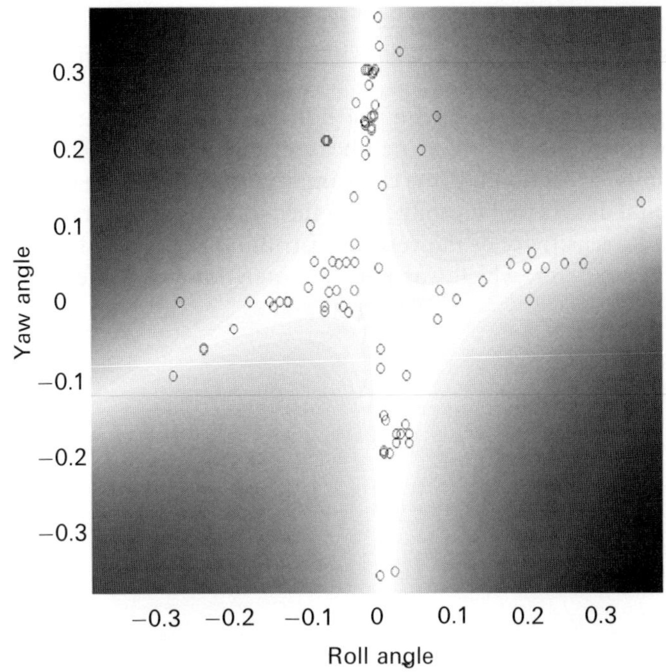

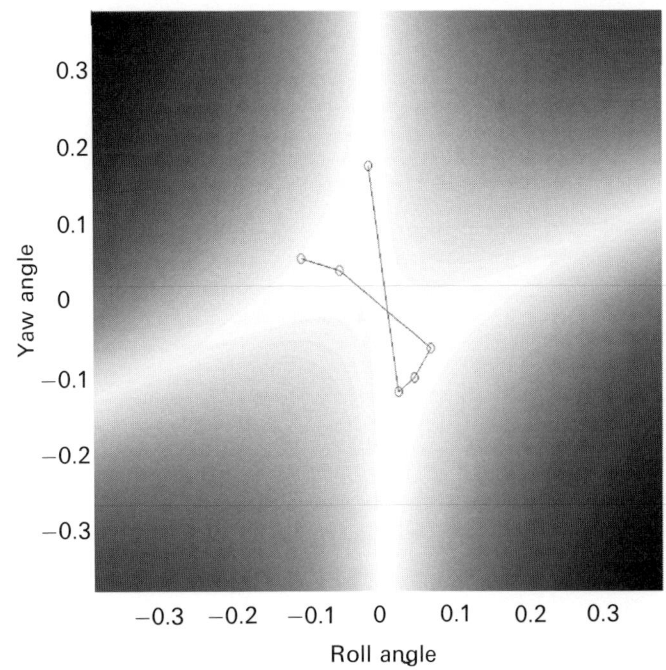

Figure 14.7. Convergence history of parameter search: (top) GA method; (bottom) DE method.

In Figure 14.8 the model-predicted received echoes obtained by the scatterer parameters shown in Table 14.3 are shown together with the experimentally observed specular echoes.

By comparing Figure 14.8 with the bottom frame of Figure 14.5, parameter inversion is seen to improve the match between the modelled and the experimental data significantly. Most notably, the echo amplitude as a function of receiver depth in the experimental data is well reproduced by the models after inversion. This improvement reflects the sensitivity of the vertical distribution of the scattered energy to the rotation of the scatterer, which was not accurately known at the outset.

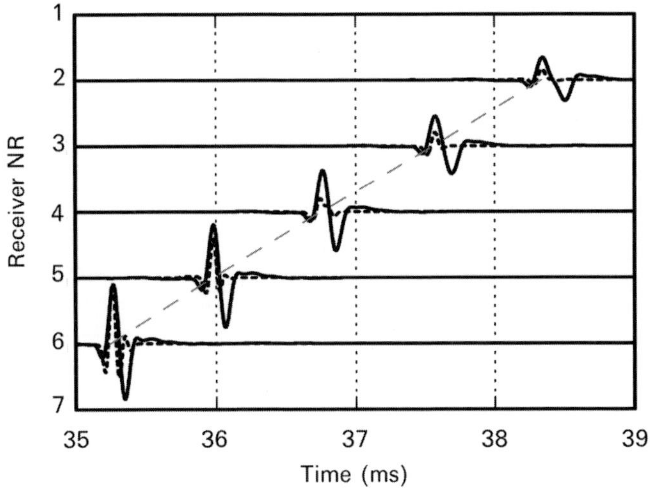

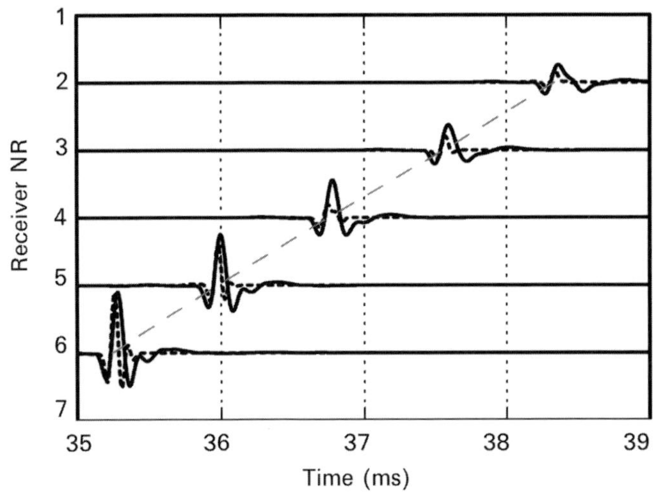

Figure 14.8. Received signals: (dotted) experimental data; (solid) modelled by parameters from the acoustic inversion shown in the right column of Table 14.3. Top: RK. Bottom: BIE.

15

Results of the biotoxicity measurements

T. Hansson, G. Åkerman, U. Tjärnlund, K. Grunder, Y. Zebühr, H. Sundberg and L. Balk

15.1 RESULTS

The results of the biotoxicity measurement are presented in Figures 15.1–15.14. In the figures the sites of the SITAR study area are arranged from north to south (left to right), and the sites within the dumping area (Möja Söderfjärd 3–7) are marked in black and white.

Neither embryonic mortality (Figure 15.1), larval mortality (Figure 15.2), deformities (Figure 15.3), nor, consequently, the sum of mortality and deformities (Figure 15.4) showed any gradient within the SITAR study area. Growth, expressed as length, was not retarded (Figure 15.5). Neither was there any increase in the dumping area for haemorrhages (Figure 15.6), heart sac oedema (Figure 15.7), yolk sac oedema (Figure 15.8), vertebral deformities (Figure 15.9) nor other morphological disorders (Figure 15.10), such as precipitate or alterations in pigmentation. In fact, there was no significant response to any of these 10 variables at any station within the SITAR study area.

The response was low also for the sub-lethal biochemical variables. Figure 15.11 shows ethoxyresorufin-*O*-deethylase (EROD) activity. EROD activity is assumed to reflect the active amount of the enzyme cytochrome P-450 1A (CYP1A), which is induced via the cytosolic aryl hydrocarbon receptor (AHR) (Gibson and Skett, 1994; Hahn, 1998; Okey *et al.*, 1994). Many planar aromatic compounds, including well-known toxic pollutants – such as planar polychlorinated biphenyls (PCBs), polycyclic aromatic hydrocarbons (PAHs) and dioxins – are strong EROD inducers. In the SITAR study area EROD values ranged between 200 and 550% of control, entailing 100 to 450% induction. This indicated an elevated baseline toxicity level in the SITAR study area compared with pristine areas. There was, however, no indication of increased toxicity in the dumping area compared with the neighbouring sites. In the Stockholm Archipelago gradient there was a substantial gradual increase in

EROD activity from Möja Söderfjärd and Möja Västerfjärd towards central Stockholm (Waldemarsudde), and a smaller but significant increase towards the open Baltic Sea (Figure 15.11). Also, the EROD activity induced by the dicyclic aromatic fraction and the polycyclic aromatic fraction in experiment 3 increased towards Stockholm (Figure 15.12). Again, the lowest potential toxicity was found in the central dumping area (Möja Söderfjärd 5). The ability of the polycyclic aromatic fraction to induce EROD activity was 3–11 times higher than that of the dicyclic aromatic fraction along the entire gradient.

Figure 15.13 shows the results of the measurement of ethoxycoumarin-*O*-deethylase (ECOD) activity, which is assumed to reflect the active amount of the enzyme cytochrome P-450 2B (CYP2B). Contaminants causing CYP2B induction are predominantly different from those causing CYP1A induction – for example, non-planar PCBs (Balk, 1985). Also, ECOD induction mechanisms are different from those governing EROD activity (Balk, 1985). ECOD values ranged between 120 and 180% of control, entailing 20 to 80% induction. As for EROD activity, this indicated a somewhat elevated baseline toxicity level in the SITAR study area compared with pristine areas, and there was no indication of increased toxicity at the dumping area compared with the neighbouring sites.

The results of the acetylcholinesterase (AChE) activity measurements are shown in Figure 15.14. AChE activity reflects the active amount of the enzyme acetylcholinesterase. Inhibition of this enzyme is a measure of neurological impairment. In other investigations AChE inhibition has been used to measure the toxicity of, for example, organophosphorus and carbamate pesticides (Galgani and Bocquené, 2000; Brewer *et al.*, 2001). The extracts did not significantly inhibit AChE activity in muscle tissue at any of the investigated sites in the dumping area, the near surroundings or the Stockholm Archipelago gradient.

The results of quantification of xenobiotics are presented in Tables 15.1 and 15.2, and Figure 15.15, which show gradually increasing sediment dry-weight concentrations from 2.5 μg ΣPCB/kg and 0.48 mg ΣPAH/kg in the central dumping area to 143 μg ΣPCB/kg and 18 mg ΣPAH/kg at Waldemarsudde in central Stockholm. Several PCBs and PAHs also showed a clear tendency towards increasing concentrations from the central dumping area towards the outer archipelago and the open sea. Injected concentrations of the analysed PCBs and PAHs are presented in Tables 15.3 and 15.4. These are compared with concentrations of PCBs and PAHs found in fish eggs from polluted areas of Lake Ontario (PCBs and PAHs, Fitzsimons *et al.*, 1995) and Lake Erie (PCBs, Russel *et al.*, 1999). The concentrations of PAHs injected into eggs exposed to sediment extracts from central Stockholm were similar to the concentrations found in eggs from a polluted area in Lake Ontario (Table 15.4). The injected concentrations of PCBs were, however, much lower than those found in eggs from polluted areas in both Lake Ontario and Lake Erie (Table 15.3). In this study we did not try to mimic the actual exposure situation for feral fish at the different sampling stations. Instead, the exposure level was chosen to facilitate comparison of potential toxicity between the investigated stations, especially with respect to the munitions dumping area. Nevertheless, it can be concluded that the exposure level in this study can indeed be reached in polluted areas.

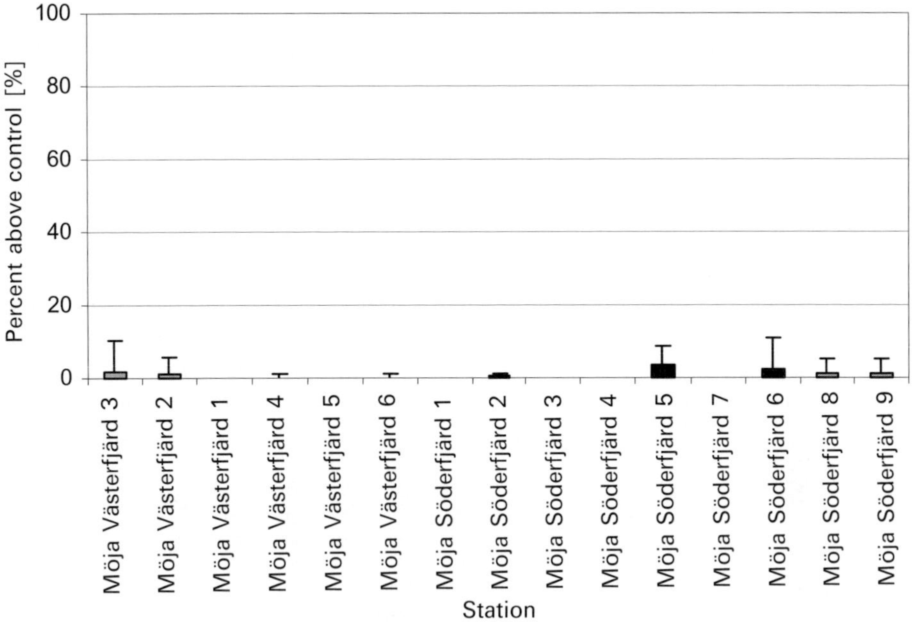

Figure 15.1. Embryonic mortality induced by the total extract. Black bars: dumping area. Error bars: 95% confidence intervals.

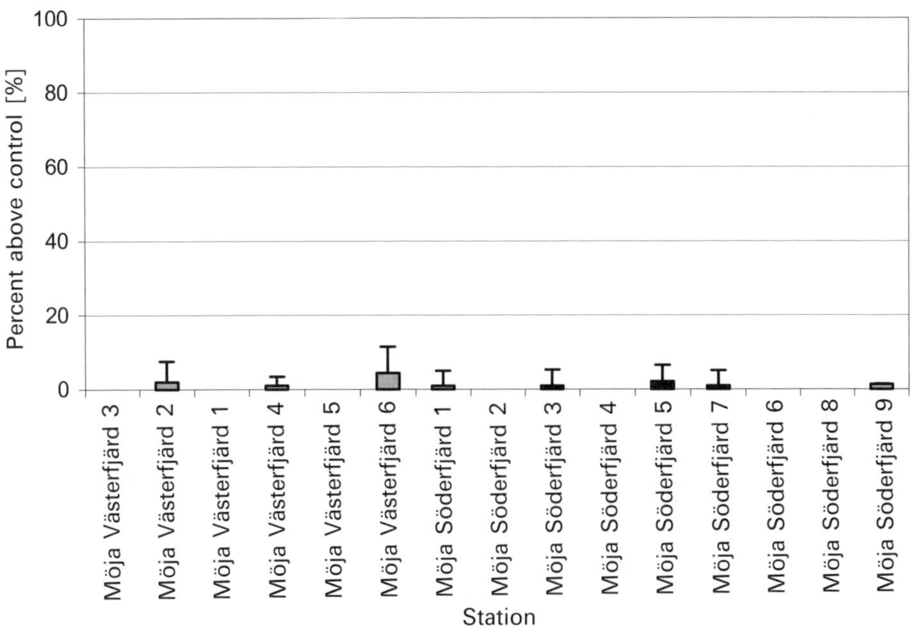

Figure 15.2. Larval mortality induced by the total extract. Black bars: dumping area. Error bars: 95% confidence intervals.

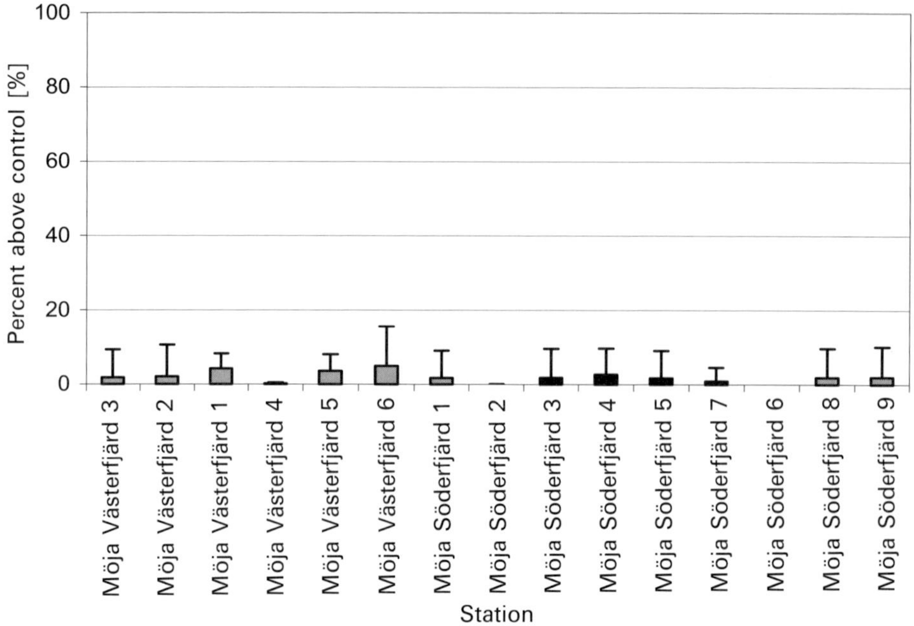

Figure 15.3. Deformities induced by the total extract. Black bars: dumping area. Error bars: 95% confidence intervals.

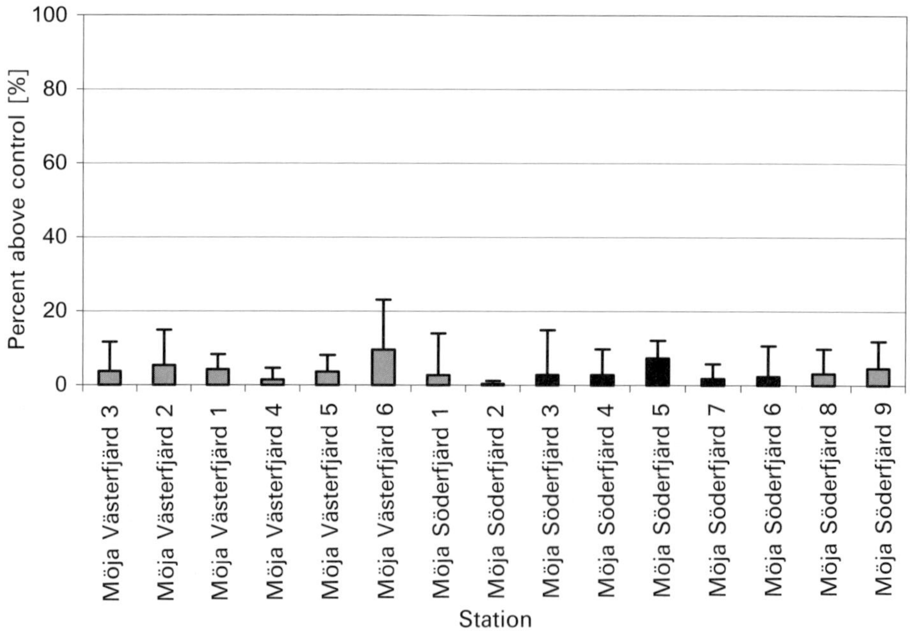

Figure 15.4. Mortality and deformities induced by the total extract. Black bars: dumping area. Error bars: 95% confidence intervals.

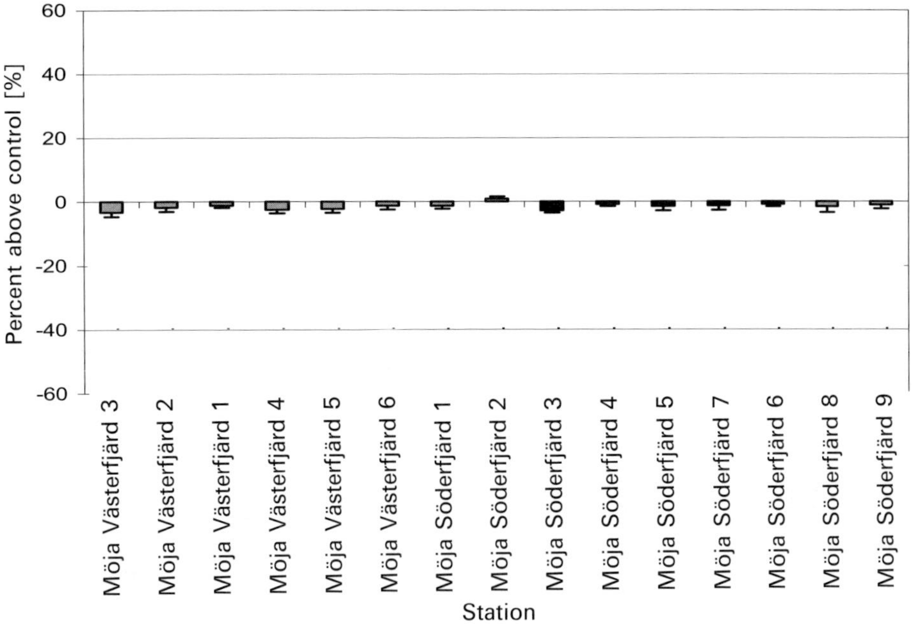

Figure 15.5. Length of larvae exposed to the total extract. Black bars: dumping area. Error bars: 95% confidence intervals. Negative length values denote retarded growth.

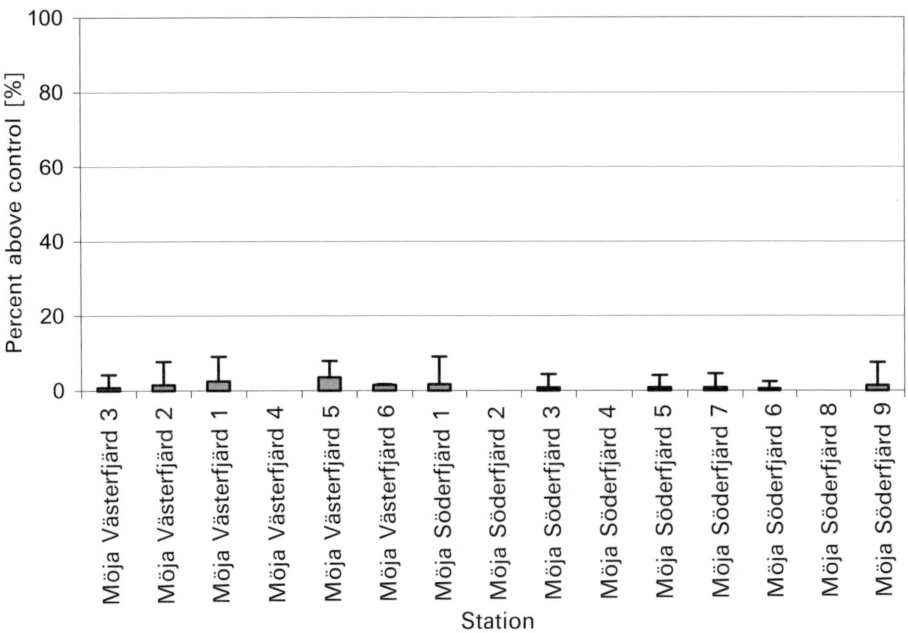

Figure 15.6. Haemorrhages induced by the total extract. Black bars: dumping area. Error bars: 95% confidence intervals.

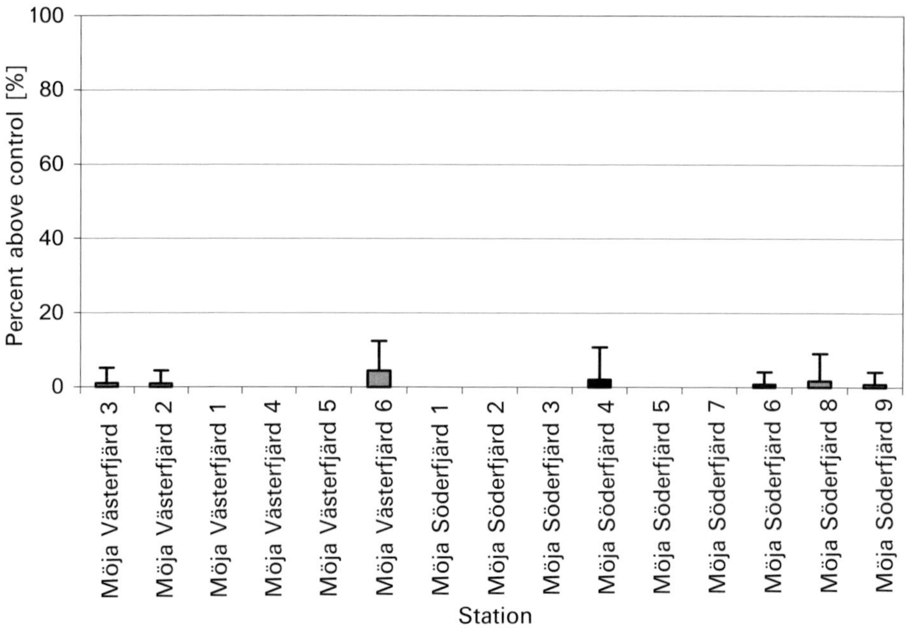

Figure 15.7. Heart sac oedema induced by the total extract. Black bars: dumping area. Error bars: 95% confidence intervals.

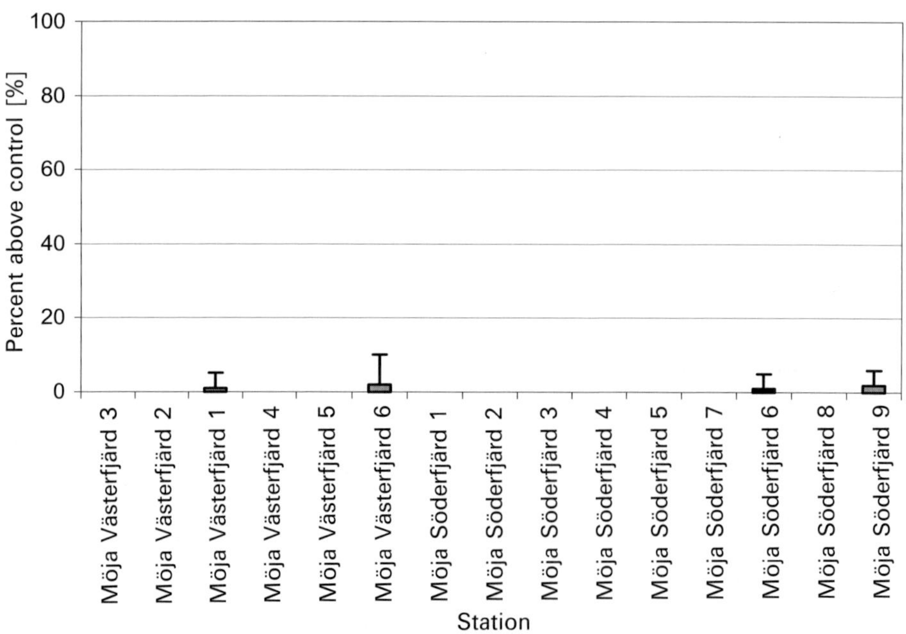

Figure 15.8. Yolk sac oedema induced by the total extract. Black bars: dumping area. Error bars: 95% confidence intervals.

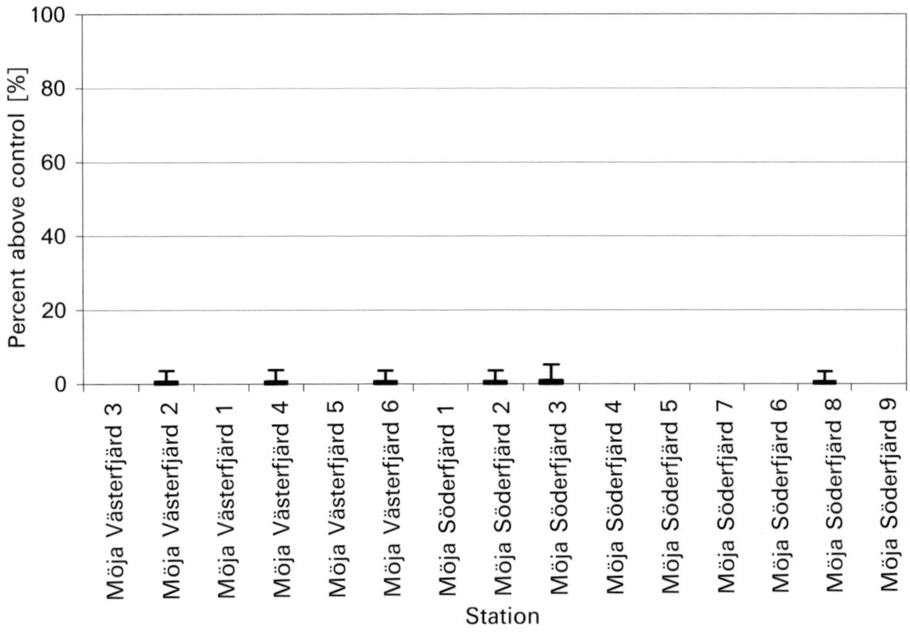

Figure 15.9. Vertebral deformities induced by the total extract. Black bars: dumping area. Error bars: 95% confidence intervals.

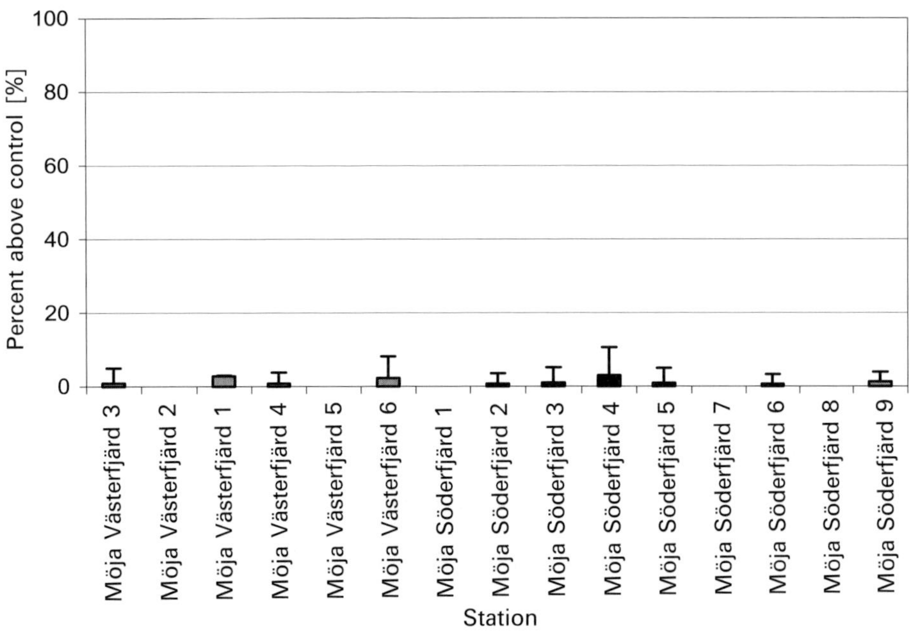

Figure 15.10. Other morphological disorders, such as precipitate or alterations in pigmentation, induced by the total extract. Black bars: dumping area. Error bars: 95% confidence intervals.

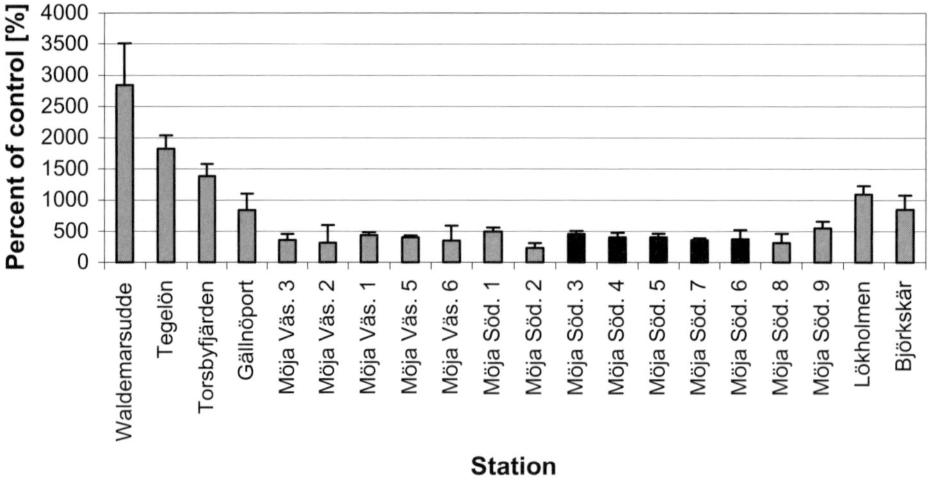

Figure 15.11. EROD activity induced by the total extract. Black bars: dumping area. Error bars: 95% confidence interval.

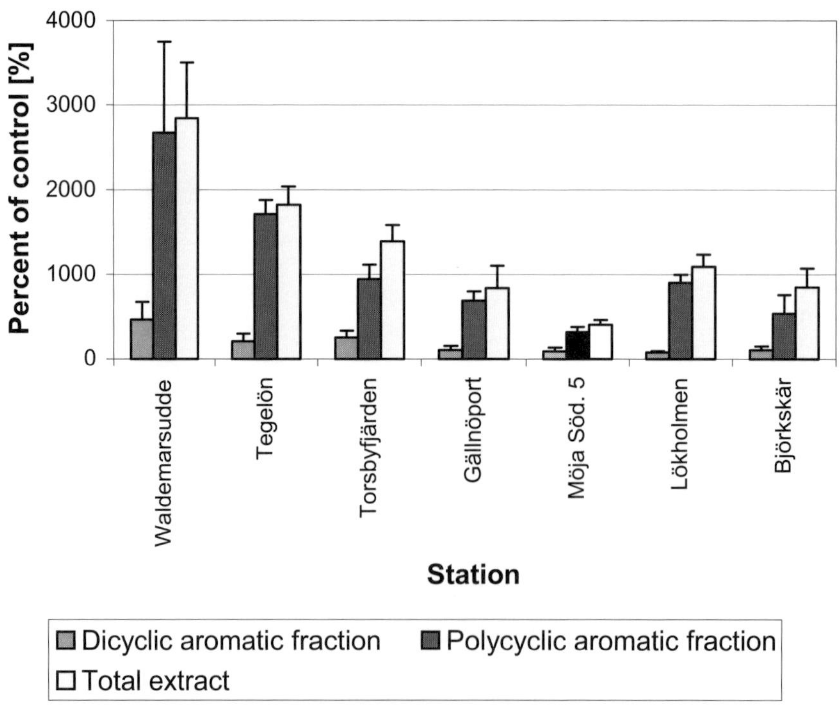

Figure 15.12. EROD activity induced by the fractions and the total extract. Grey/black/white bars: dumping area. Error bars: 95% confidence interval.

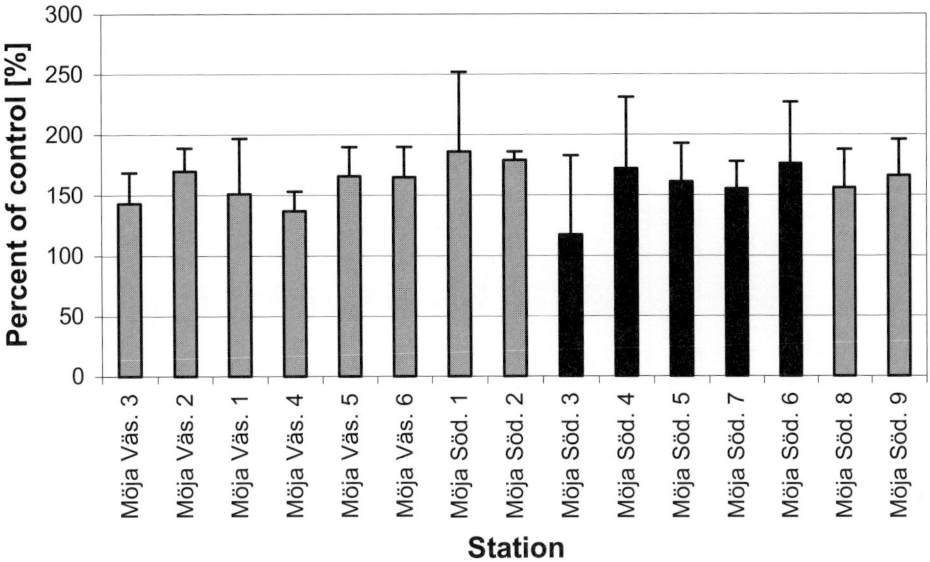

Figure 15.13. ECOD activity induced by the total extract. Black bars: dumping area. Error bars: 95% confidence interval.

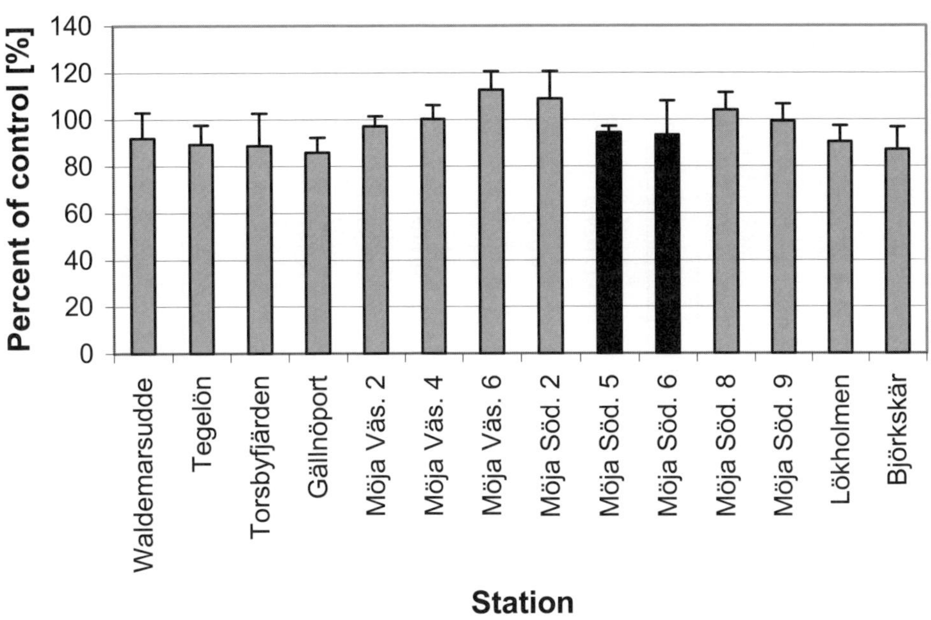

Figure 15.14. AChE activity in muscle tissue of larvae exposed to the total extract. Black bars: dumping area. Error bars: 95% confidence interval.

Table 15.1. Concentration of polychlorinated biphenyls (PCBs) in sediment along the Stockholm Archipelago gradient (ng/kg dry sediment). Accuracy: 2 significant digits.

	Waldemarsudde	Tegelön	Torsbyfjärden	Gällnöport	Möja Söderfjärd 5	Lökholmen	Björkskär
PCB 18	4 845	1 681	202	75	47	25	42
PCB 28	17 029	6 054	1 910	343	155	83	158
PCB 52	8 440	3 094	1 590	285	88	109	129
PCB 70	9 057	2 415	919	281	125	109	161
PCB 90/101	14 743	5 686	3 350	648	203	284	309
PCB 105	4 178	1 853	877	239	63	140	130
PCB 110	13 637	5 367	2 884	503	195	240	235
PCB 118	10 369	4 404	2 823	653	185	308	327
PCB 138	17 404	9 479	4 962	1 395	345	547	620
PCB 149	14 099	5 841	4 042	854	155	375	302
PCB 153	16 442	7 870	5 300	1 281	727	536	615
PCB 180/193	10 740	4 046	2 654	724	171	309	312
PCB 194	1 771	714	410	100	30	60	57
PCB 199	174	48	46	9	6	10	6
ΣPCB	142 929	58 552	31 967	7 390	2 495	3 133	3 401

Table 15.2. Concentration of polycyclic aromatic hydrocarbons (PAHs) in sediment along the Stockholm Archipelago gradient [µg/kg dry sediment]. Accuracy: 2 significant digits.

	Waldemarsudde	Tegelön	Torsbyfjärden	Gällnöport	Möja Söderfjärd 5	Lökholmen	Björkskär
Phenanthrene	958	123	78	38	21	23	24
3-Methyl-phenanthrene	172	21	15	7	4	3	4
1-Methyl-phenanthrene	185	23	14	6	5	4	5
Fluoranthene	1 594	253	209	128	51	75	64
Pyrene	1 850	288	189	45	23	31	50
2-Phenyl-naphthalene	90	12	11	6	3	4	4
Benzo[a]fluoranthene	390	51	37	12	6	9	7
Benzo[e]fluoranthene	416	56	34	10	3	6	8
2-Methyl-pyrene	187	25	18	6	2	3	4
1-Methyl-pyrene	132	19	12	5	2	3	3
Benzo[ghi]fluoranthene	373	63	56	37	8	17	21
Cyclopenta[cd]pyrene	995	117	85	37	9	14	23
Benzo[a]anthracene	2 110	245	187	70	19	31	48
Chrysene+Triphenylene	1 854	196	183	105	22	52	44
2,2′-Binaphthalene	29	5	5	4	0	2	2
Benzo[b]fluoranthene	1 665	301	262	216	79	119	109
Benzo[j+k]fluoranthene	696	107	99	76	36	37	42
Benzo[e]pyrene	865	154	130	78	33	45	45
Benzo[a]pyrene	1 223	176	140	60	19	37	38
Perylene	377	80	92	77	33	33	55
Indeno[1,2,3-cd]pyrene	915	193	171	162	57	74	69
Benzo[ghi]perylene	815	203	170	110	43	60	40
Coronene	45	30	24	6	3	6	2
ΣPAH	17 935	2 742	2 223	1 301	482	689	712

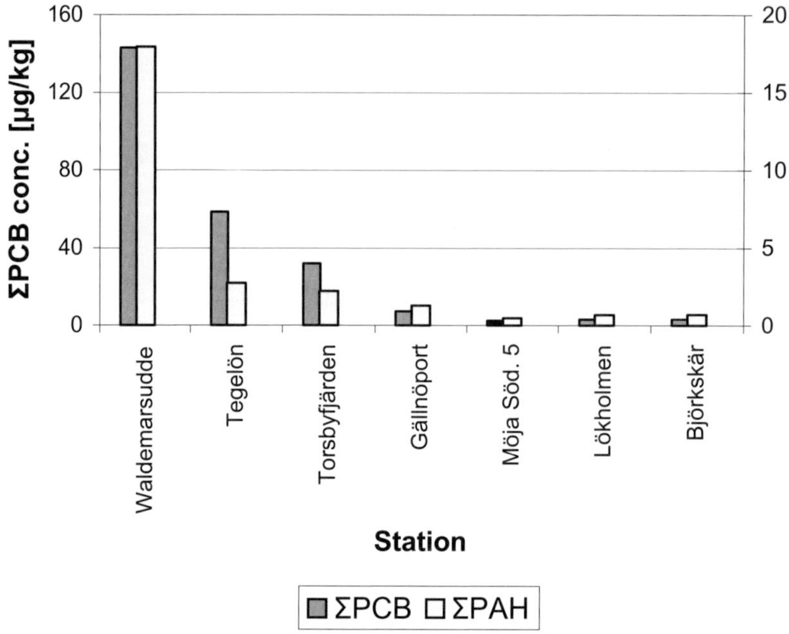

Figure 15.15. ΣPCB and ΣPAH concentrations in dry sediment along the Stockholm Archipelago gradient. The analysed PCB congeners are listed in Table 15.1 and the analysed PAHs are listed in Table 15.2. Black/white bars: dumping area.

15.2 DISCUSSION

Since none of the investigated variables showed a significant response in the dumping area or its near surroundings, the conclusion was that no measurable lipophilic toxic substances were released from the dumped munitions. The conclusion is valid for at least 2000–2002, since that is the period when the investigated sediment was formed. It should be emphasized, though, that the lack of toxicity was shown only for this relatively small dumping area and that the results cannot be extrapolated to other dumping areas without reservations, as there may be differences both in the amount and the quality of the dumped munitions. It should also be mentioned that not all types of toxicity were measured in this investigation. For example, endocrine disorders and immunotoxicity were not investigated.

There was, not surprisingly, a clear co-variation between ΣPCB, ΣPAH and EROD activity. At Möja Söderfjärd and Möja Västerfjärd the injected concentrations of the analysed PCBs and PAHs were too low to be responsible for more than a small part of EROD induction. Many other pollutants, known to coexist with the analysed PCBs and PAHs, may however account for the observed EROD induction. The results thus indicate significant background pollution in Möja Söderfjärd and Möja Västerfjärd. At Waldemarsudde in central Stockholm the analysed PCBs and PAHs were present in sufficient concentrations to explain, at least partly, the EROD

Table 15.3. Injected concentrations of polychlorinated biphenyls (PCBs) in rainbow trout (*Oncorhynchus mykiss*) eggs exposed to sediments from the Stockholm Archipelago gradient (Waldemarsudde–Björkskär) compared with measured mean PCB concentrations in eggs from feral fish in Lake Ontario[a] (lake trout, *Salvelinus namaycush*) and Lake Erie[b] (whitefish, *Coregonus clupeaformis* and black crappie, *Pomoxis nigromaculatus*) (ng/kg egg ww). Accuracy: 2 significant digits.

	Waldemar-sudde	Tegelön	Torsby-fjärden	Gällnö-port	Möja Söderfjärd 5	Lökholmen	Björkskär	Lake Ontario[a]	Lake Erie[b]	
	Rainbow trout							Lake trout	White-fish	Black crappie
PCB 18	60.57	21.01	2.52	0.94	0.58	0.31	0.53			
PCB 28	212.86	75.67	23.87	4.29	1.93	1.04	1.97	10 000	960	1 690
PCB 52	105.50	38.67	19.87	3.56	1.10	1.36	1.61	29 000	4 750	7 598
PCB 70	113.22	30.19	11.49	3.52	1.57	1.36	2.02		3 789	6 803
PCB 90/101	184.29	71.08	41.88	8.10	2.54	3.55	3.86	77 000	7 820	20 937
PCB 105	52.23	23.17	10.97	2.98	0.79	1.75	1.63	27	2 312	7 322
PCB 110	170.46	67.08	36.05	6.29	2.43	3.00	2.94		5 517	14 025
PCB 118	129.61	55.05	35.29	8.16	2.31	3.84	4.09	44 000		15 917
PCB 138	217.55	118.49	62.03	17.44	4.31	6.84	7.75	95 000	10 804	37 165
PCB 149	176.24	73.01	50.52	10.68	1.94	4.68	3.77		7 652	19 848
PCB 153	205.53	98.38	66.25	16.01	9.09	6.70	7.68	104 000	8 734	33 277
PCB 180/193	134.24	50.57	33.17	9.05	2.14	3.87	3.90	50 000	5 121	20 003
PCB 194	22.14	8.93	5.12	1.25	0.37	0.75	0.71		831	2 985
PCB 199	2.17	0.60	0.57	0.11	0.08	0.13	0.07			
ΣPCB	1 786.61	731.91	399.59	92.38	31.19	39.17	42.52			

[a] Data from Fitzsimons *et al.*, 1995.
[b] Data from Russel *et al.*, 1999.

Table 15.4. Injected concentrations of polycyclic aromatic hydrocarbons (PAHs) in rainbow trout (*Oncorhynchus mykiss*) eggs exposed to sediments from the Stockholm Archipelago gradient (Waldemarsudde–Björkskär) compared with measured mean PAH concentrations[a] in eggs from feral fish in Lake Ontario[b] (lake trout, *Salvelinus namaycush*) (μg/kg egg ww). Accuracy: 2 significant digits.

	Waldemar-sudde	Tegelön	Torsbyfjärden	Gällnöport	Möja Söderfjärd 5	Lökholmen	Björkskär	Lake Ontario[a,b]
	Rainbow trout							Lake trout
Phenanthrene	11.98	1.54	0.97	0.47	0.26	0.29	0.30	17.90
3-Methyl-phenanthrene	2.16	0.27	0.19	0.09	0.05	0.03	0.05	
1-Methyl-phenanthrene	2.31	0.28	0.18	0.07	0.06	0.05	0.06	
Fluoranthene	19.93	3.16	2.61	1.60	0.64	0.93	0.80	12.20
Pyrene	23.12	3.60	2.36	0.57	0.29	0.38	0.63	498.60
2-Phenyl-naphthalene	1.12	0.15	0.14	0.08	0.04	0.05	0.05	
Benzo[a]fluoranthene	4.88	0.63	0.47	0.15	0.08	0.12	0.09	
Benzo[e]fluoranthene	5.20	0.70	0.42	0.13	0.04	0.08	0.10	
2-Methyl-pyrene	2.33	0.31	0.23	0.07	0.02	0.04	0.05	
1-Methyl-pyrene	1.65	0.24	0.16	0.06	0.03	0.04	0.04	
Benzo[ghi]fluoranthene	4.66	0.78	0.70	0.47	0.10	0.21	0.27	
Cyclopenta[cd]pyrene	12.44	1.46	1.06	0.46	0.11	0.17	0.28	
Benzo[a]anthracene	26.37	3.07	2.34	0.87	0.24	0.38	0.60	
Chrysene+Triphenylene	23.18	2.45	2.29	1.31	0.27	0.65	0.54	
2,2′-Binaphthalene	0.36	0.06	0.06	0.05	0.00	0.03	0.03	
Benzo[b]fluoranthene	20.81	3.76	3.28	2.70	0.98	1.49	1.36	10.30
Benzo[j+k]fluoranthene	8.70	1.34	1.24	0.96	0.45	0.47	0.53	
Benzo[e]pyrene	10.82	1.93	1.63	0.98	0.41	0.56	0.56	
Benzo[a]pyrene	15.29	2.20	1.75	0.75	0.24	0.46	0.48	11.20
Perylene	4.71	1.00	1.15	0.96	0.41	0.41	0.69	
Indeno[1,2,3-cd]pyrene	11.43	2.42	2.14	2.02	0.71	0.93	0.86	12.00
Benzo[ghi]perylene	10.18	2.54	2.13	1.38	0.54	0.75	0.50	
Coronene	0.57	0.38	0.30	0.08	0.04	0.08	0.02	
ΣPAH	224.18	34.28	27.79	16.26	6.02	8.61	8.89	

[a] The mean was based on samples with detectable levels.
[b] Data from Fitzsimons *et al.*, 1995.

inducing ability of the injected sediment extracts. Although it is beyond the aim of these investigations, the results suggest a severe pollution situation in central Stockholm. The insensitivity of AChE activity to the extract of this polluted sediment is worthy of note.

The sum of the EROD activity of the dicyclic and polycyclic aromatic fractions was nearly equal to the EROD activity of the total extract, indicating that no potent EROD inducers other than those in the dicyclic and polycyclic aromatic fractions were present. The observation that the polycyclic aromatic fraction was by far the most potent is in agreement with several other investigations (Brunström *et al.*, 1992; Sundberg *et al.*, 2005a, b).

16

3D acoustical image analysis and feature extraction

A. Caiti, V. Murino, M. Palmese, A. Trucco and M. Zakharia

The processing chain introduced in Chapter 6 was successfully applied to analyse some 3D acoustic images, obtained by both simulated and real signals acquired by different sonar systems, and containing objects that were completely or partially buried. The availability of simulated data provided the opportunity of evaluating the performances of the developed methods on the basis of objective measures of errors. Using field data from the tank experiment it was verified that the measurements extracted from the final 3D model were close to the real ones, according to the resolution limits of the employed sonar systems. As explained in Chapter 14, the MAS survey in the SITAR study area did not provide sufficiently dense data to allow application of the image analysis and feature extraction tools of Chapter 6.

16.1 RESULTS ON SIMULATED DATA

Data generated by the simulation method described in Palmese and Trucco (2004) offer the opportunity of estimating the robustness of the image processing chain described in Chapter 6. A sonar configuration has been assumed that is based on an acoustic transmitter, ensonifying all the volume of interest by a pulse 40 μs long with a carrier frequency of 50 kHz, and a square receiving array composed of 20×20 hydrophones, placed 2 m above the bottom and parallel to the seafloor. The volume examined was $115 \times 115 \times 61$ cm^3 wide, where a longitudinal slice of a cylinder, buried a few tens of centimetres, was placed (see Figure 16.1). Data provided by such a geometrical configuration are presented in Figures 16.2–16.3 (all in colour section). Each 3D image was defined as a regular grid of cubic voxels[1] with a side of length 1 cm; each voxel was assigned an intensity value ranging from 0 to 255. From the slices

[1] A voxel is a volume element, representing a value on a regular grid in 3D space (similar to the pixel, or picture element, in 2D space).

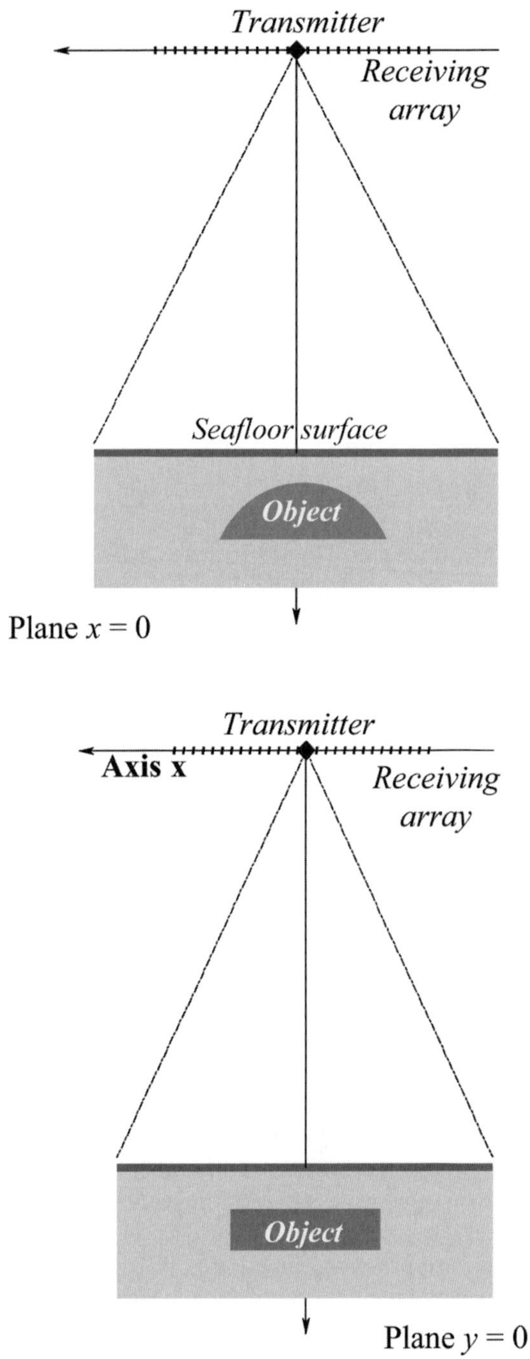

Figure 16.1. Geometry of the simulated scene: longitudinal slice of a 20-cm buried cylinder; projections on the planes $x = 0$ and $y = 0$.

displayed in Figures 16.2–16.3, it was possible to distinguish the acoustic responses of the surface of the seafloor, of the volumetric sediment and of the buried object with linear and dB visualization, respectively. Figure 16.4(a) (see colour section) shows the same volume slice as in Figure 16.3(b) after decreasing the resolution by two levels in an octree[2] structure (i.e., by a factor of 64 in data reduction); noise reduction in the image is clearly visible as well as the effect of the lower resolution. Figure 16.4(b) shows – again for the slice corresponding to Figure 16.3(b) – the results obtained by applying a $3 \times 3 \times 3$ median filter: speckle reduction, less granularity in sediment reverberation and smoother object reverberation can clearly be noticed. In both cases, it should be emphasized that, although the results are shown for 2D slices, all the processing was applied through 3D implementations.

The segmentation results (on raw data, with no pre-filtering) are presented in Figures 16.5–16.6, showing the satisfactory and robust performance of the method. Starting with 3D raw data, both in linear and dB scale, we tested the segmentation procedure comparing the different performances corresponding to the following aggregation criteria:

(*i*) mean intensity criterion;
(*ii*) statistical approach based on *a priori* information;
(*iii*) statistical approach based on the use of the Weibull density function.

In Figure 16.5 we display the statistical segmentation results obtained on the linear raw data when the PDFs chosen for the regions to be segmented (object, seafloor surface and sediment) were assumed to be Gaussian for object and seafloor surface scattering, and Rayleigh for sediment reverberation. Two different resolution-level representations are shown: Figure 16.5(b) visualizes the second level of the octree structure of the segmented volume image, thus providing a coarse but still effective rendering, whereas Figure 16.5(a) shows the segmented object at the highest resolution. Starting with the same volumetric data – but using a logarithmic scale for the intensity levels – the segmentation results obtained with the mean intensity criterion and with the statistical approach based on the use of the Weibull density function are shown in Figures 16.6(a) and 16.6(b), respectively.

The availability of *a priori* information makes it possible to obtain the best segmentation results in terms of accuracy; nevertheless, if this information is not accessible, the use of the Weibull PDF (which is able to change its shape according to current data distributions) jointly with LSM for parameter estimation allows one to achieve a good trade-off between segmentation performance and processing speed, whereas the mean intensity criterion provides poorer results, but in a very short time. These considerations are supported by evaluation of the degree of under-segmentation and over-segmentation of the different results. Starting with the same volumetric raw data and the same seeds we compared the level of under-segmentation of the different approaches by evaluating the number of segmented voxels in excess of those in the synthetic model and the level of over-segmentation by estimating the

[2] An octree is a multi-resolution pyramid for representation of 3D spaces (see full explanation in Section 6.3).

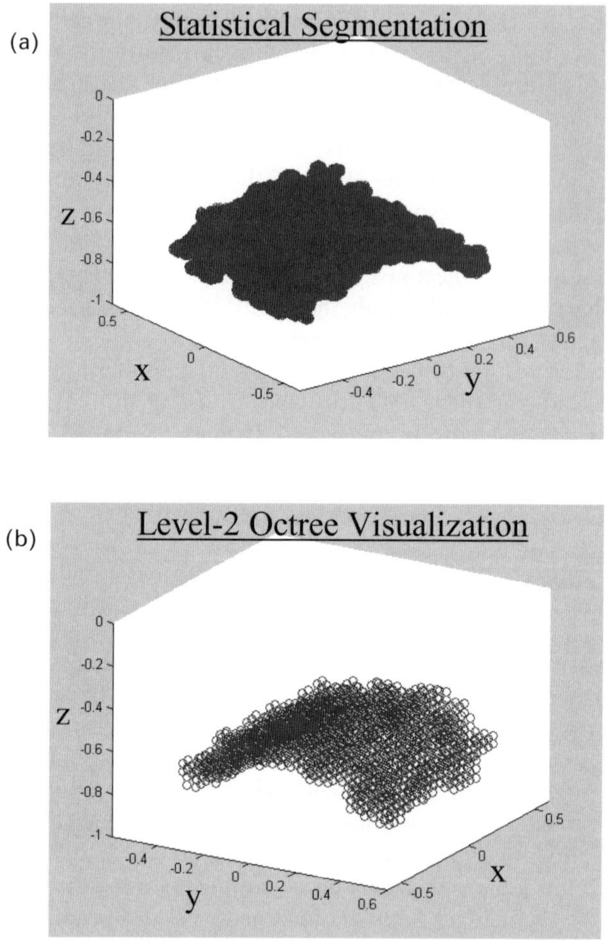

Figure 16.5. Segmentation results: (a) highest-resolution visualization; (b) octree level-2 image representation.

extent of gaps in the object surface. The degrees of under- and over-segmentation were computed by averaging the results of many segmentation processes (each one being characterized by a different choice of seeds). The 3D histograms of Figures 16.7(a) and 16.7(b) show such measures as percentages and their relative standard deviations for the following segmentation criteria:

(1) mean intensity (MI);
(2) *a priori* statistical approach – Gaussian PDF for the object and the seafloor surface, and Rayleigh PDF for the sediment – (G-G-R);
(3) Weibull statistical approach (W);
(4) Gaussian statistical approach – when the PDFs of all classes are forced to be Gaussian – (G).

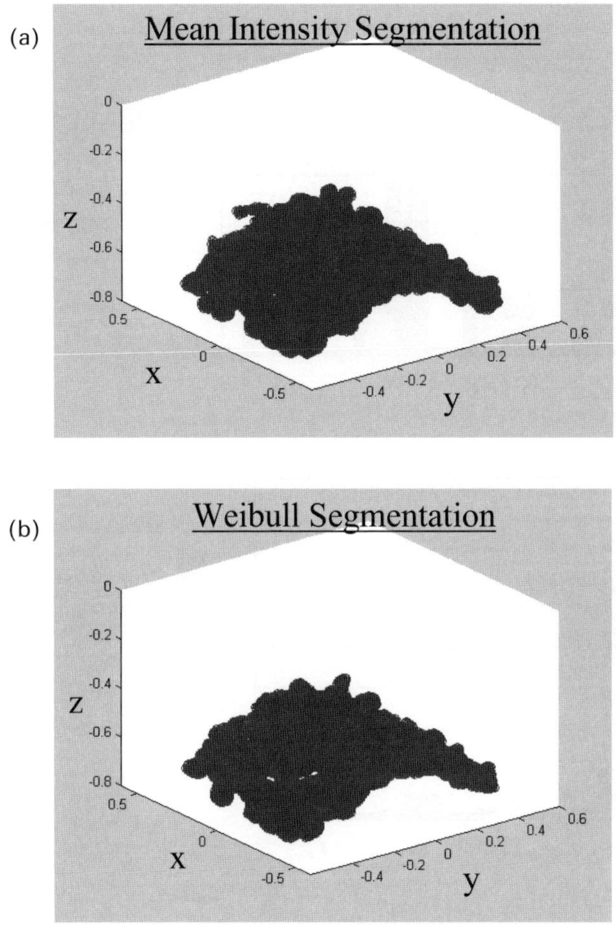

Figure 16.6. Segmentation results: (a) mean intensity criterion; (b) Weibull statistical criterion.

From analysis of the histograms presented in Figure 16.7, it is possible to note that the statistical approach based on *a priori* information is the most reliable, as it not only provides lower mean degrees of under- and over-segmentation, but the relative standard deviations in both cases are the smallest.

By analysing Figure 16.5(a) it is possible to note that the segmented object preserves the properties of symmetry of the simulated cylinder allowing us to apply directly the inertial tensor method to estimate the pose of the target. In Figure 16.8 we show the values of the parameters extracted starting with the segmented data of Figure 16.5(a) – that is:

(1) the coordinates of the barycentre with respect to the inertial coordinate system;
(2) orientation by means of the Euler angles (θ, φ);

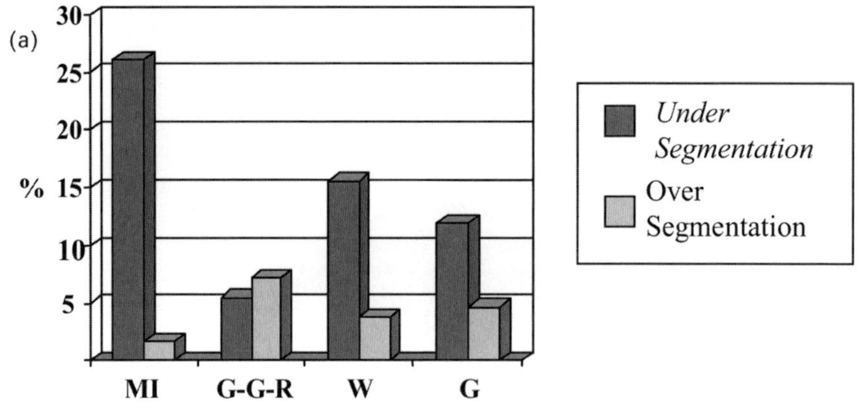

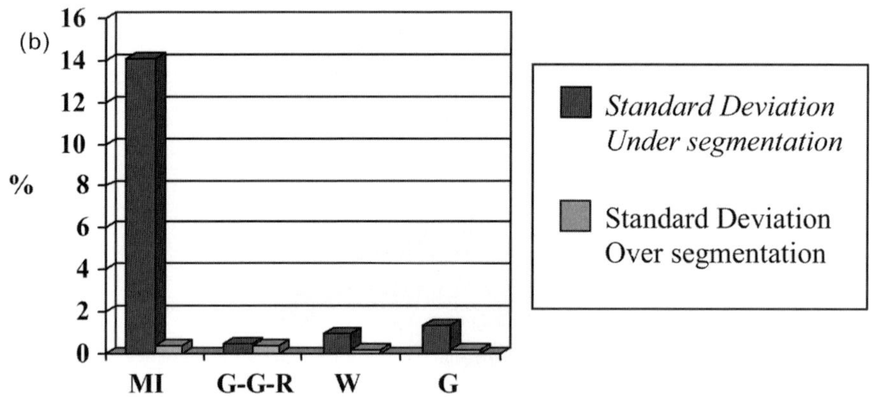

Figure 16.7. Segmentation performance evaluation: (a) under- and over-segmentation measures; and (b) measurement of standard deviation.

(3) the radius of the cylinder;
(4) the height of the cylinder;
(5) the volume of the whole cylinder.

Finally, distribution of the points belonging to the segmented object and the cylinder obtained by fitting the points are visualized in Figure 16.9.

To evaluate the errors affecting the extracted measures it is necessary to focus on the sonar system performance in terms of spatial resolution. The angular resolution r_a of the simulated system can be approximated by the following formula (Murino and Trucco, 2000):

$$r_a = 2 \cdot 0.44 \cdot \frac{\lambda}{N \cdot d} = 0.88 \cdot \frac{c}{N \cdot d \cdot f} \qquad (16.1)$$

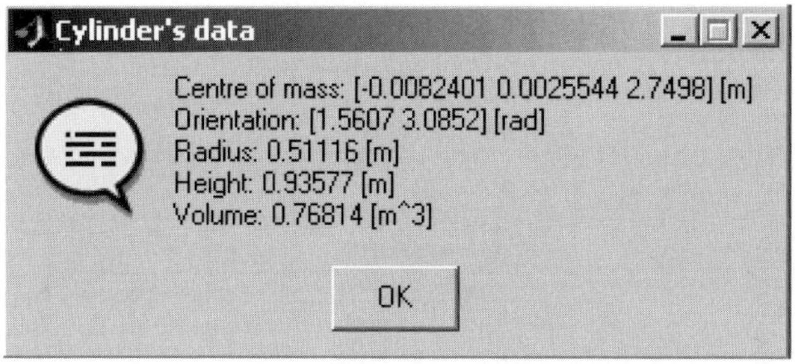

Figure 16.8. Features extracted by the segmented data of Figure 16.5(a).

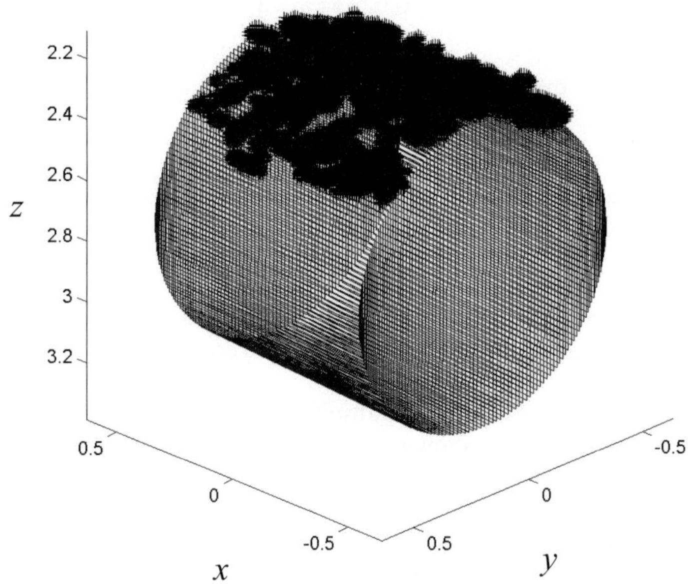

Figure 16.9. Segmented data and related, fitted, cylinder 3D visualization.

where λ is the wavelength of the emitted signal, f is the carrier frequency, N is the number of sensors on one side of the square array, d is the inter-element distance and c is the sound velocity in the medium.

A measure of the range resolution r_r is given by the following expression:

$$r_r = \frac{T \cdot c}{2} \tag{16.2}$$

where T is the temporal length of the pulse transmitted by the sonar. The sonar system configuration used for the considered simulation is characterized by the following parameters:

$$\left.\begin{aligned} c &= 1477\ \mathrm{m/s} \\ f &= 50\ \mathrm{kHz} \\ N &= 20 \\ d &= 0.06\ \mathrm{m} \\ T &= 40\ \mu\mathrm{s} \end{aligned}\right\} \qquad (16.3)$$

On the basis of these values, the expressions of Equations (16.1) and (16.2) provide the angular and range resolutions of the system, respectively:

$$\left.\begin{aligned} r_a &= 0.021\,66\ \mathrm{rad} \\ r_r &= 0.029\,54\ \mathrm{m} \end{aligned}\right\} \qquad (16.4)$$

Moreover, as the simulated scene reaches a maximum depth p of 2.5 m, we can obtain the maximum value for the lateral resolution by multiplying such depth by the angular resolution r_a:

$$r_l \cong r_a p = 0.054\,15\ \mathrm{m} \qquad (16.5)$$

We can now compare our estimated measures with the highest value between range and lateral resolution:

$$res = \max\{r_l, r_r\} = 0.054\,15\ \mathrm{m} \qquad (16.6)$$

In Table 16.1 we have listed the estimated features versus the values set at the beginning of the simulation and the relative absolute errors which must be evaluated taking into account the tolerance values.

It can be noted in general that all the absolute errors are lower or comparable with their relative tolerance value, showing the good performances of the developed methods. Let us focus our attention on the different parameters:

- Barycentre coordinates: their tolerance was set equal to the system resolution, *res*.
- Radius and height: their tolerance was set equal to 2*res*, as the computation of these parameters involves estimation of the positions of two points (extreme points of a segment) each affected by an error at least equal to *res*.

Table 16.1. Comparison between the estimated features and the simulation values set.

	Estimated value	Simulated value	Absolute error	Tolerance
Barycentre x [m]	−0.008 240 1	0	0.008 240 1	0.054 15
Barycentre y [m]	0.002 554 4	0	0.002 554 4	0.054 15
Barycentre z [m]	2.749 8	2.7	0.049 8	0.054 15
θ [rad]	1.560 7	1.570 8	0.010 8	0.135 3
φ [rad]	3.085 2	3.141 6	0.056 4	0.135 3
Radius [m]	0.511 16	0.5	0.011 16	0.108 3
Height [m]	0.935 77	0.8	0.135 77	0.108 3

- Euler angles: their tolerance was computed taking into account the maximum linear error produced by incorrect estimation of the inclination. For each angle we computed the displacement of the furthest voxel with respect to the barycentre belonging to the object and placed in the direction identified by the considered angle. This displacement must be compared with sonar system resolution. In the table we have inserted the equivalent tolerance value expressed in radians.

Starting from radius and height estimations, the volume of the entire cylinder and the relative percentage error can be computed. In particular, the real and estimated volumes, vol_{obj} and vol_{est}, are equal to $0.62\,\mathrm{m}^3$ and $0.76\,\mathrm{m}^3$, respectively, and the relative percentage error, $e_{\%vol}$, is equal to 22%:

$$e_{\%vol} = \frac{|vol_{obj} - vol_{est}|}{vol_{obj}} = \frac{(0.76 - 0.62)}{0.62} = 22\% \tag{16.7}$$

Finally, concerning VRML visualization, the point distribution representing the segmented region related to the target was described by means of a set of coordinates (x_i, y_i, z_i) with respect to the inertial system and characterized by the same level. In Figures 16.10 and 16.11 we have shown the same scene observed by three different points of view and zooms chosen by the user. In Figure 16.10 the scene is visualized from the default point of view and zoom: the object, which is the segmented longitudinal slice of an empty cylinder, is rendered as a cloud of black points, whereas the water volume and sediments are represented as dark and light grey regions, respectively. Starting with the same virtual world representing the segmentation result, it is

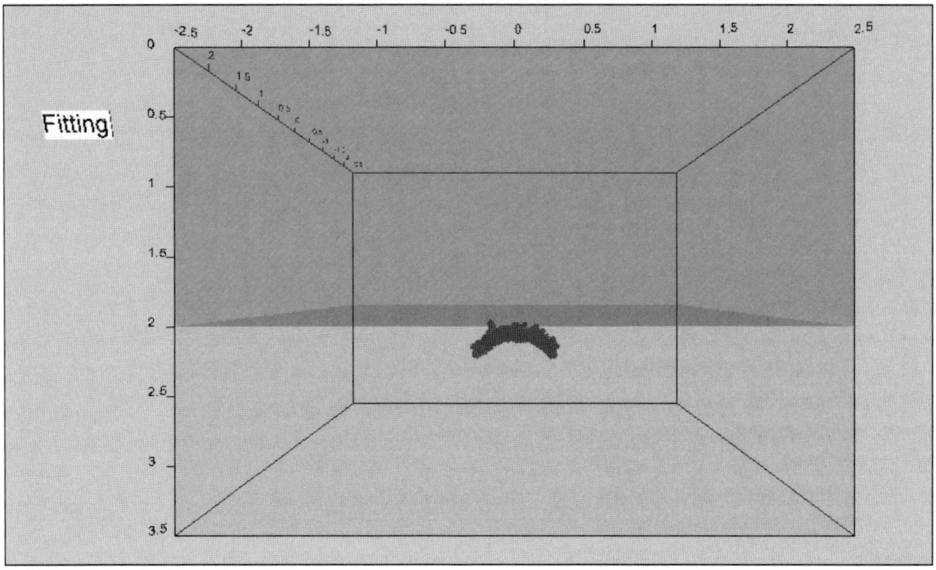

Figure 16.10. Segmentation result visualized by VRML.

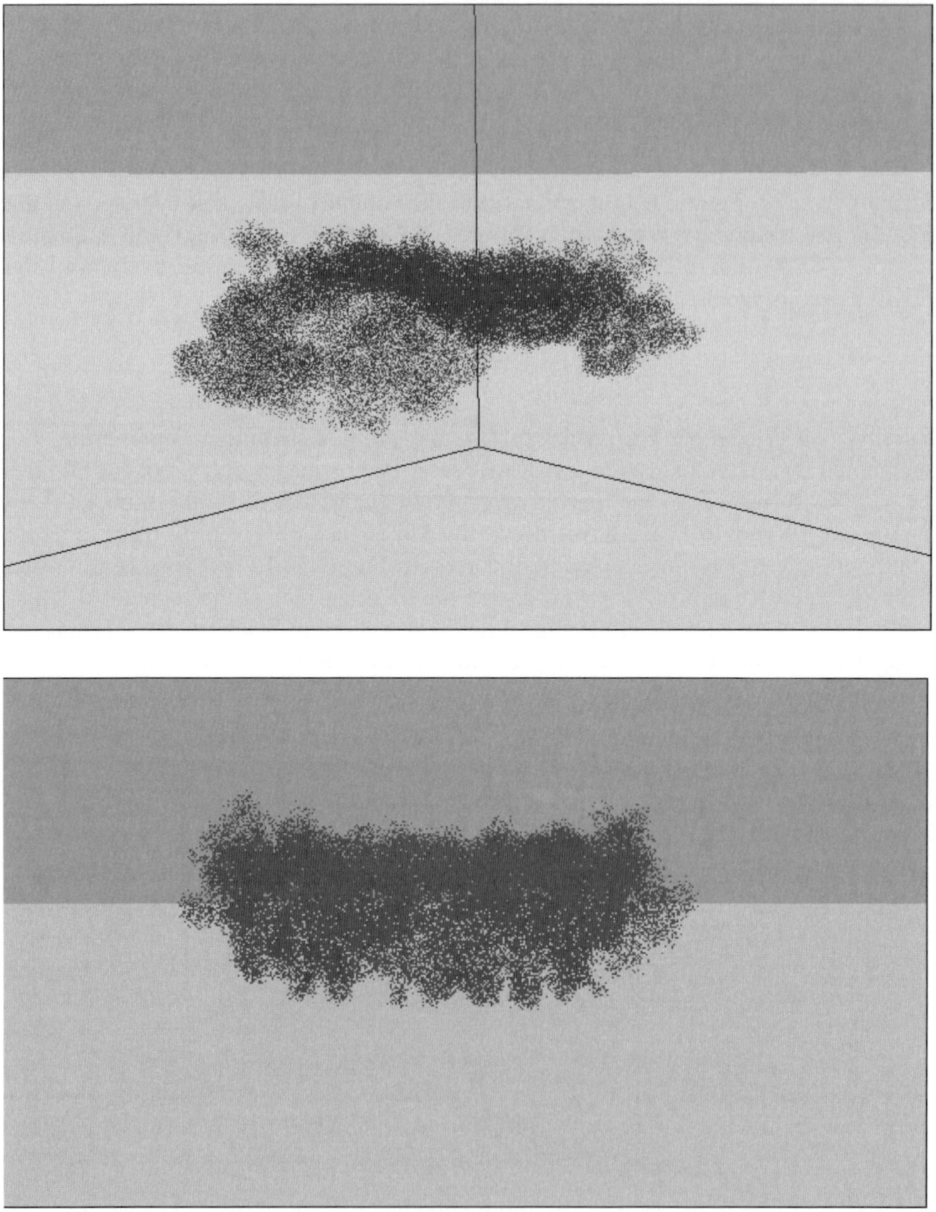

Figure 16.11. VRML visualization of the segmentation result from different points of view chosen by the operator.

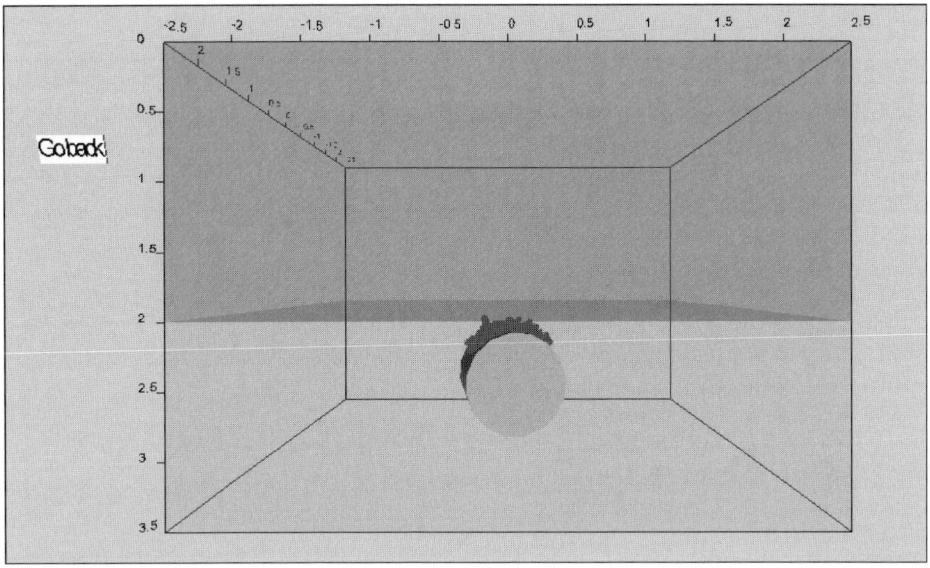

Figure 16.12. Segmentation result and virtual reconstructed cylinder visualized by VRML.

possible to add the reconstructed cylinder by simply mouse-clicking on the button "Fitting" (see Figure 16.10). In Figures 16.12 and 16.13 we show the same scene represented in Figures 16.10 and 16.11 but with the fitted cylinder visualized.

16.2 RESULTS ON EXPERIMENTAL DATA

In order to further validate the developed methods, real data coming from two different-scaled tank experiments were also analysed. Of the two experiments, the first is not part of the SITAR activities, but was made available to the SITAR partners by Manell Zakharia, ARMINES/EN.

The first experiment set-up is depicted in Figure 16.14. The experiment was carried out in a water tank 2 m long, 1 m wide and 1 m deep. The target was a cylinder buried in a sand layer. 3D raw data were obtained by a planar synthetic aperture imaging technique, where the transmitted signal was a tone burst at 1.1 MHz. Figure 16.15 (see colour section) shows three slices of the volumetric raw data; from the side views the scattering strength corresponding to the seafloor surface and to the buried object are clearly visible. In particular, from Figure 16.15(c) it is possible to note a pronounced depression of the bottom sediment in the tank that corresponds to the cylinder position.

Before presenting the results obtained on this set of data it is useful to give some information about the system resolution. The range resolution δ_r of the planar

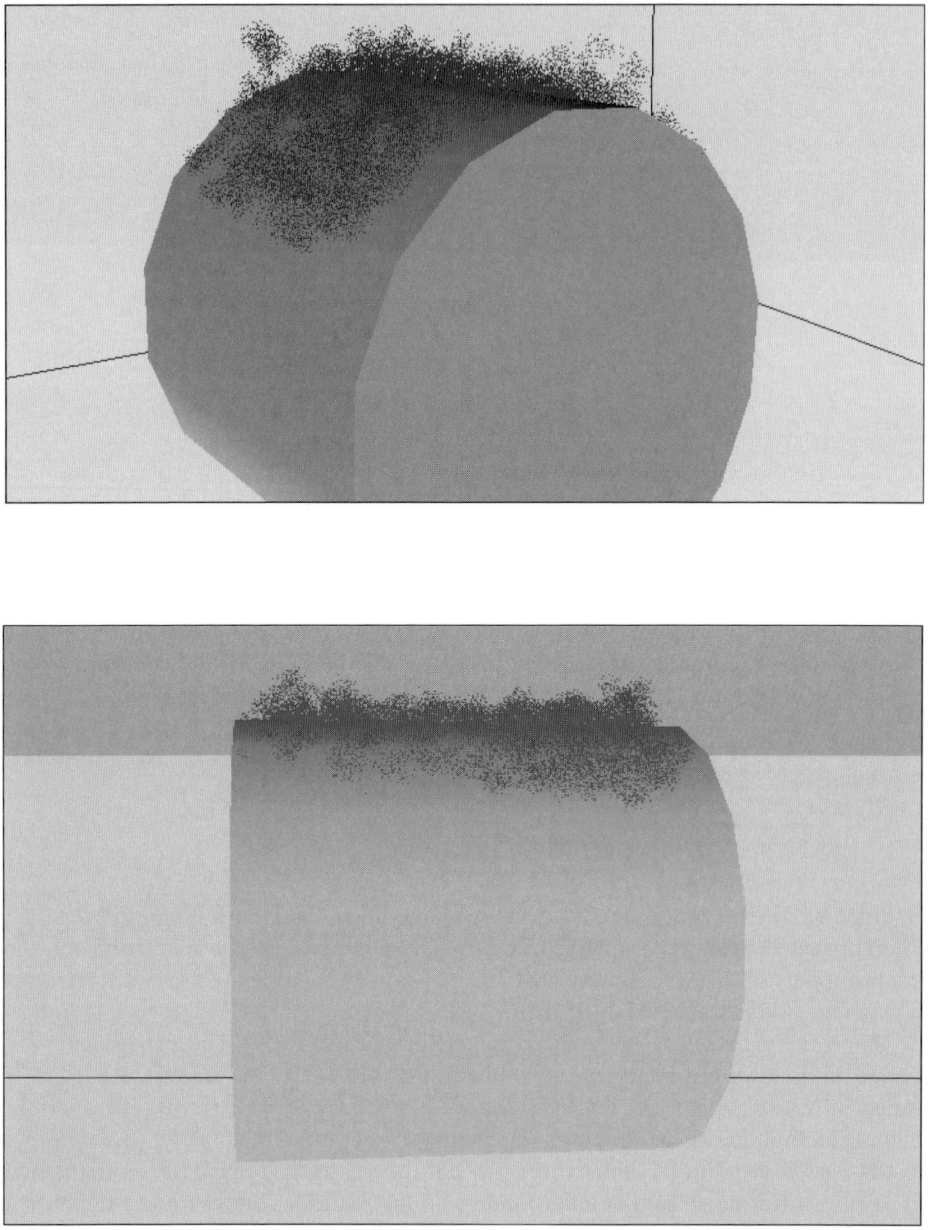

Figure 16.13. Segmentation results and related virtual reconstructed cylinders from different points of view.

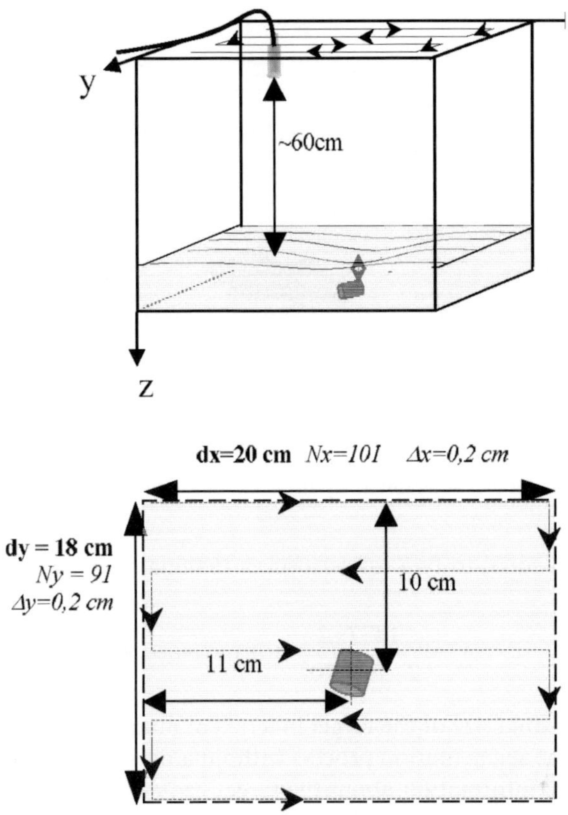

Figure 16.14. Geometry of the first tank experiment: the target was a buried cylinder.

synthetic aperture sonar (P-SAS) was the same as that achieved by conventional sonar systems, and so can be expressed by the following relation:

$$\delta_r = \frac{c_0 \cdot T}{2} \tag{16.8}$$

where, again, c_0 is the sound velocity in water and T the temporal length of the transmitted signal. In contrast, the lateral resolution δ_{sa} provided by P-SAS was better than that characteristic of traditional sonar. For both the azimuth and elevation directions the P-SAS processing was able to obtain lateral resolution independent of the distance between the sonar system and the considered point of the scene. In particular, δ_{sa} is only related to the diameter ϕ of the transducer:

$$\delta_{sa} = \frac{\phi}{2} = 0.31 \text{ cm} \tag{16.9}$$

where the P-SAS transducer diameter has been set equal to $\phi \cong 0.62$ cm according to the experimental set-up. By considering the sound velocity equal to $c_0 = 1477$ m/s and

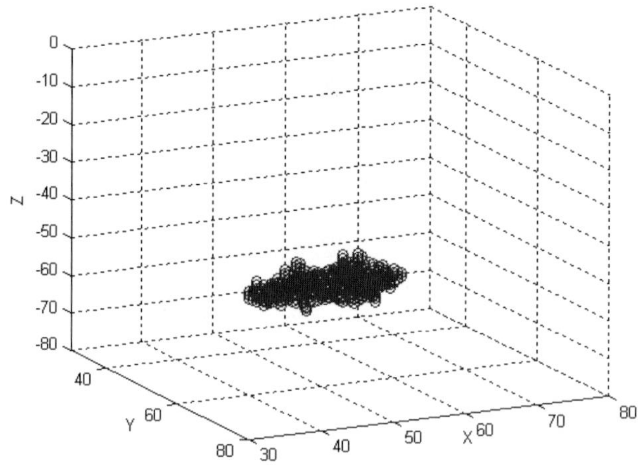

Figure 16.16. Object segmented by the statistical volume growing algorithm.

the impulse length equal to two carrier cycles, the final value assumed by the range resolution is:

$$\delta_r = \frac{1477 \cdot 2}{2 \cdot 1.1 \cdot 10^6} \cong 0.134 \text{ cm} < \delta_{sa} \tag{16.10}$$

Considering the limited size of the target in terms of number of voxels, it was decided to directly apply the segmentation process without a noise reduction step, in order to preserve any useful information about the target geometry. We chose the segmentation method based on the use of the Weibull PDF to exploit the best performance of an accurate statistical investigation. Figure 16.16 shows the obtained segmented object composed of 119 voxels.

The very limited number of voxels makes it difficult to extract parameters about the object. We have supposed that the object points detected by the sonar and thus segmented belong only to the target surface. This hypothesis only affects estimation of the radius and barycentre position, whereas the height and the inclination of the object are not compromised. The features extracted by the inertial tensor method and according to the cylinder model are presented in Figure 16.17.

The real values for the diameter d_{obj}, height h_{obj} and the corresponding volume vol_{obj} of the object are, respectively:

$$\left.\begin{aligned} d_{obj} &= 2.5 \text{ cm} \\ h_{obj} &= 2.6 \text{ cm} \\ vol_{obj} &= 12.76 \text{ cm}^3 \end{aligned}\right\} \tag{16.11}$$

Over-estimation of the height value can be justified considering the united effect due to the system resolution ($=0.31$ cm) and to the applied method of estimation that provides an upper limit to such a parameter. Combining such effects, the error e_t

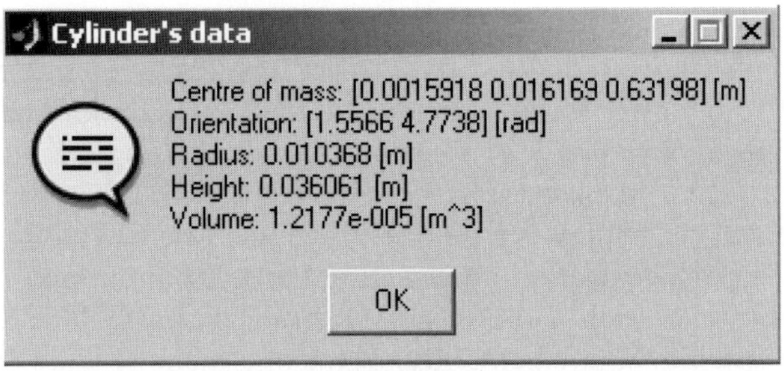

Figure 16.17. Features extracted by the segmented data of Figure 16.16.

that can be tolerated in the measurement of the cylinder height can be formulated in the following way:

$$e_t > 2 \cdot 0.31 \text{ cm} = 0.62 \text{ cm} \quad \Rightarrow \quad e_t \approx 0.7 \text{ cm} \tag{16.12}$$

which is comparable with the real committed error. All the other estimated parameters ended up being in conformity with the spatial resolution achieved by the imaging sonar system. In particular, the volume of the cylinder, vol_{est}, ended up as 12.17 cm^3; thus, the relative percentage error, $e_{\%vol}$, in the volume estimation is equal to:

$$e_{\%vol} = \frac{|vol_{obj} - vol_{est}|}{vol_{obj}} = \frac{(12.76) - 12.17)}{12.76} = 4.6\% \tag{16.13}$$

In Figure 16.18 the object reconstructed on the basis of the above hypothesis and according to the adopted cylinder model is presented together with the original segmented data.

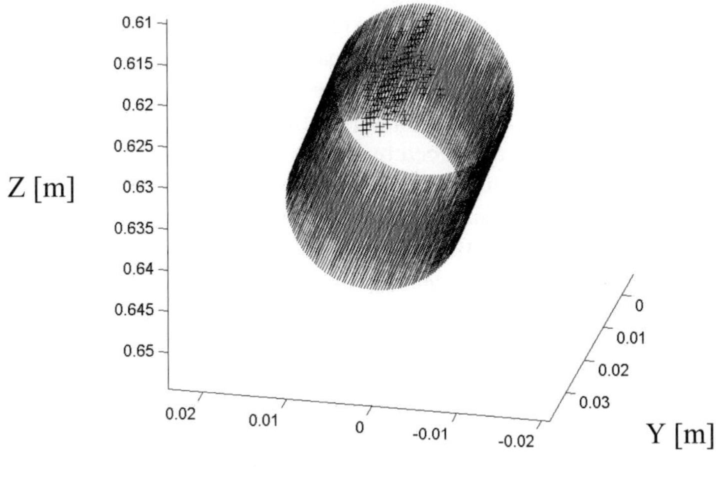

Figure 16.18. Reconstructed data: cylinder fitted on the segmented data assuming only surface scattering.

The second experiment considered comes from the MAS-scaled tank data experiment described in Chapter 8. Raw data (after processing in order to extract the scattering-relative intensity to be associated at each x–y–z coordinate, see Chapter 3) were organized in a $41 \times 41 \times 41$ matrix of voxels, where each voxel represents a cubic volume of $1\,\text{cm}^3$. In Figure 16.19 (see colour section) three slices of a 3D raw image are presented. The object is a solid steel cylinder oriented along the z-axis. Note that the object is partially buried, intersecting the seafloor surface.

A free software Matlab toolbox of 3D visualization (V3D, 2004) was partly adapted to our specific requirements. By means of a menu (Figure 16.20, see colour section) the user can inspect raw data by choosing the slice of interest (parallel to plane x, y, z) in the 3D data. Moreover, it is possible to visualize a sequence of slices by clicking on the button "Animation" of the menu. Figure 16.20 shows the toolbox interface: in panel (a) a slice parallel to plane y has been selected by the user; in panel (b) the tool menu is displayed. In this way the user can rapidly and easily obtain information about the data to be analysed.

Considering the limited size of the target in terms of the number of voxels, it was decided to directly apply the segmentation process without a noise reduction step in order to preserve any useful information about the target geometry. We chose the segmentation method based on the use of the PDF fitting to optimise performance. Figure 16.21 shows the obtained segmented object composed of 290 voxels (the data were from the same object whose raw data generated Figure 4.5.19). Starting with the segmentation results, the parameters extracted by the inertial tensor method and according to the cylinder model are presented in Figure 16.22.

The limited number of voxels made difficult the task of reconstruction and parameter extraction. Nevertheless, the estimated features were comparable with the real ones taking into account the spatial resolution of raw data and the accuracy of object positioning. In particular, relative errors of 50% in length and 2% in radius were obtained. The volume estimation error was 41%. In Figure 16.23 the object reconstructed on the basis of the extracted parameters and according to the adopted cylinder model is presented together with the original segmented data.

Finally, concerning VRML visualization, the point distribution representing the target-segmented region was described by means of a set of coordinates (x_i, y_i, z_i) with respect to the inertial system and associated with the same representation level. Figure 16.24(a) shows the segmented cylinder rendered as a cloud of black points, whereas the water volume and the sediments are represented as dark and light grey regions, respectively. By mouse-clicking on the button "Fitting" it is possible to visualize the reconstructed cylinder – see Figure 16.24(b). Once in the VRML representation, the user can zoom, translate, rotate and look at the object from different orientations.

Let us present the results on another 3D acoustic image. Figure 16.25 (see colour section) shows three slices of the raw data referring to a buried cylinder, with orientation along the x-axis.

The segmentation procedure was applied directly to raw data without any preprocessing operation. In Figure 16.26 the cloud of voxels labelled as belonging to the target is shown, confirming again the powerful potentialities of the developed

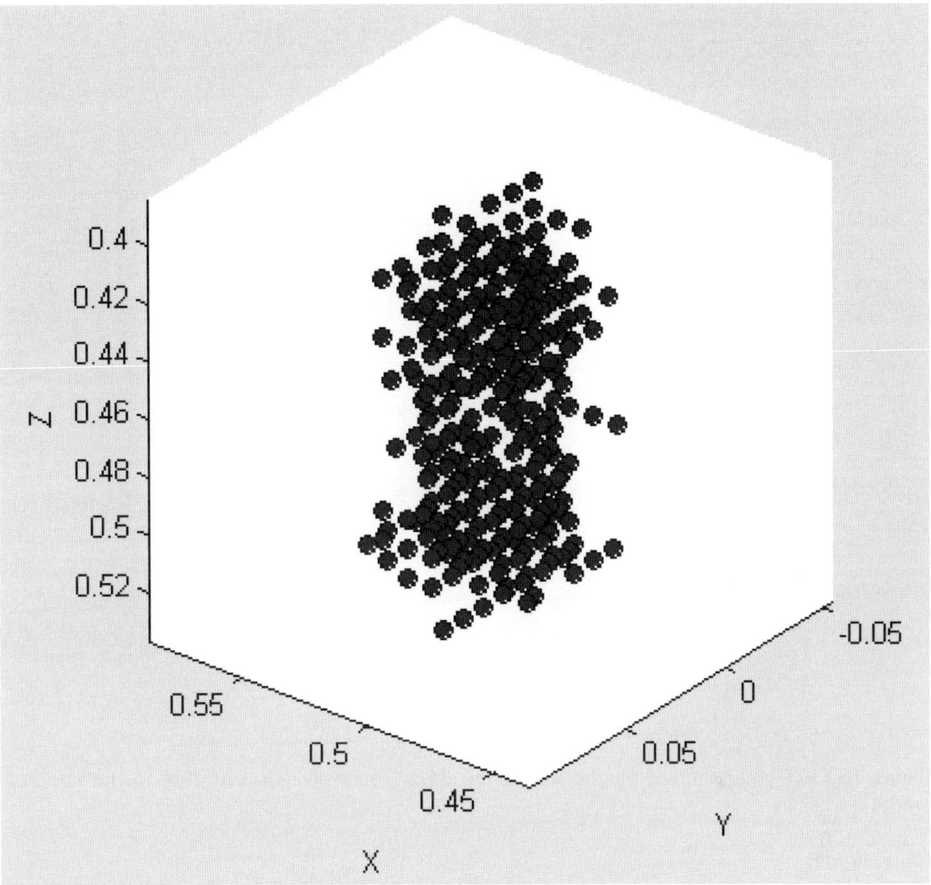

Figure 16.21. Segmentation result (segmented data for target 4) obtained by the statistical volume growing segmentation process.

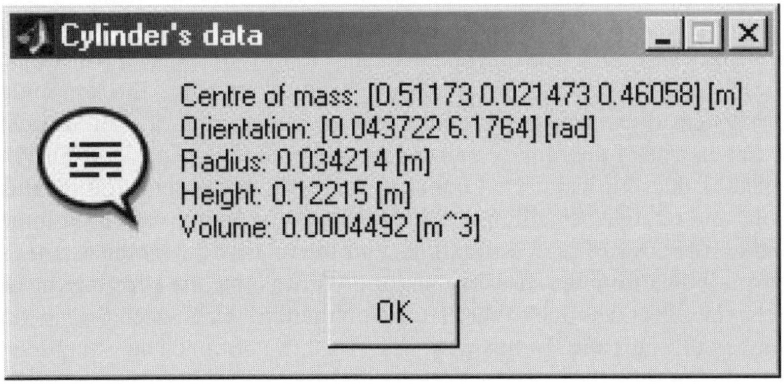

Figure 16.22. Features extracted from the segmented data of Figure 16.21.

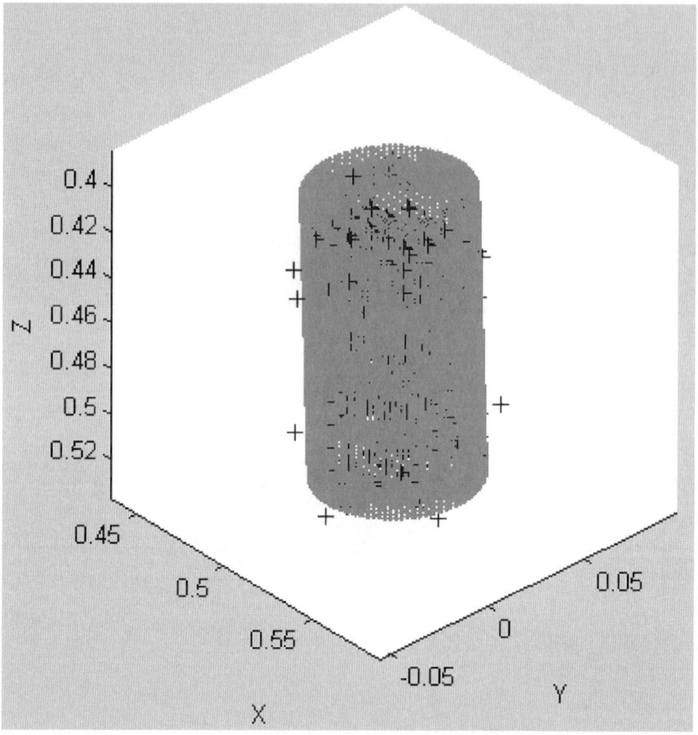

Figure 16.23. Cylinder fitted on the segmented data (for target 4) according to the adopted model.

segmentation algorithm. The features extracted and the relative reconstructed cylinder model are presented in Figures 16.27 and 16.28, respectively. The relative errors in object dimension estimation were 63% in length and 0% in radius. The estimated volume relative error was 63%.

In conclusion, taking into account all the MAS datasets acquired during the scaled tank experiment (nine of them), errors of the order of 10–30% in target radius estimation and of 40–60% in target height estimation were reported, with objects of 5–10-cm size. Considering the 10:1 scaling of the experiment, this amounted to a maximum 60-cm dimension reconstruction error in objects of 1-m size. In Table 16.2 the dimensions of the four cylindrical targets used in the considered experiments are detailed. Tables 16.3 and 16.4 contain, for each experiment, radius and height estimations and relative percentage errors, respectively. Moreover, the volume values computed on the basis of such estimations and the relative percentage errors are also shown. For all the examined cases the major error was the one affecting the length of the barrels; this effect was due to the estimation method itself which provided only an upper limit to this measure. In this way, in general, we obtained an estimated volume which was greater than the real one; even so, in the worse case the percentage error did not exceed 66%.

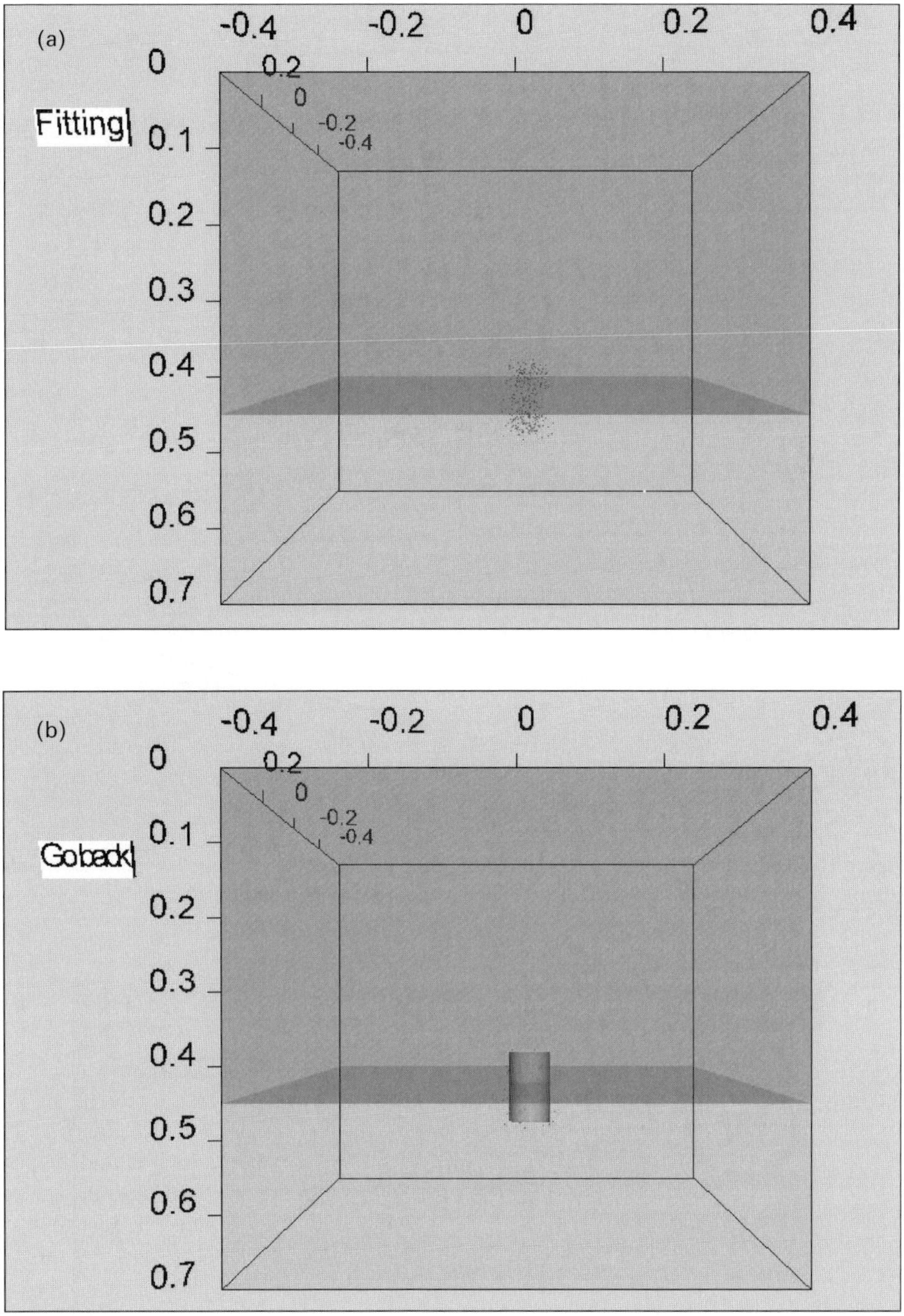

Figure 16.24. (a) VRML rendering of the segmentation result; and (b) virtual reconstruction of the target.

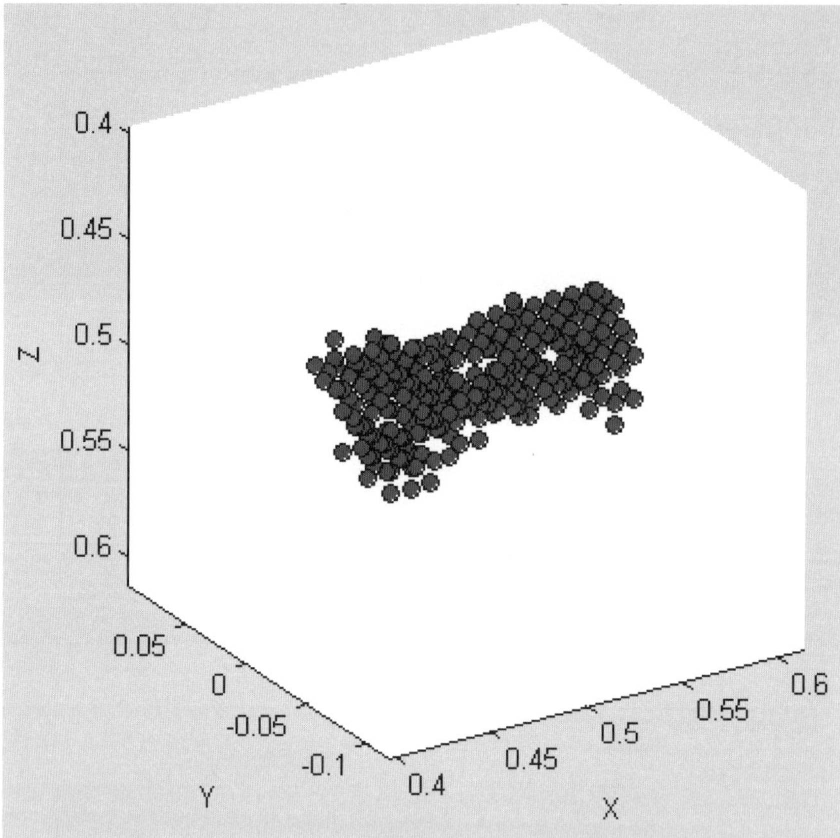

Figure 16.26. Segmentation result (segmented data for target 2) obtained by the statistical volume growing segmentation process for the same object as in Figure 16.25.

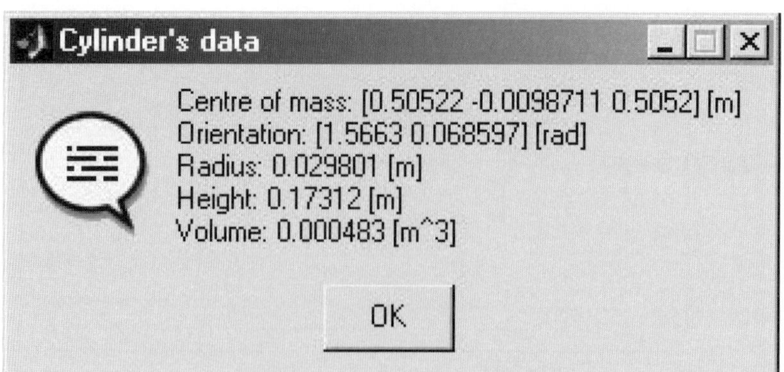

Figure 16.27. Features extracted from the segmentation distribution of Figure 16.26.

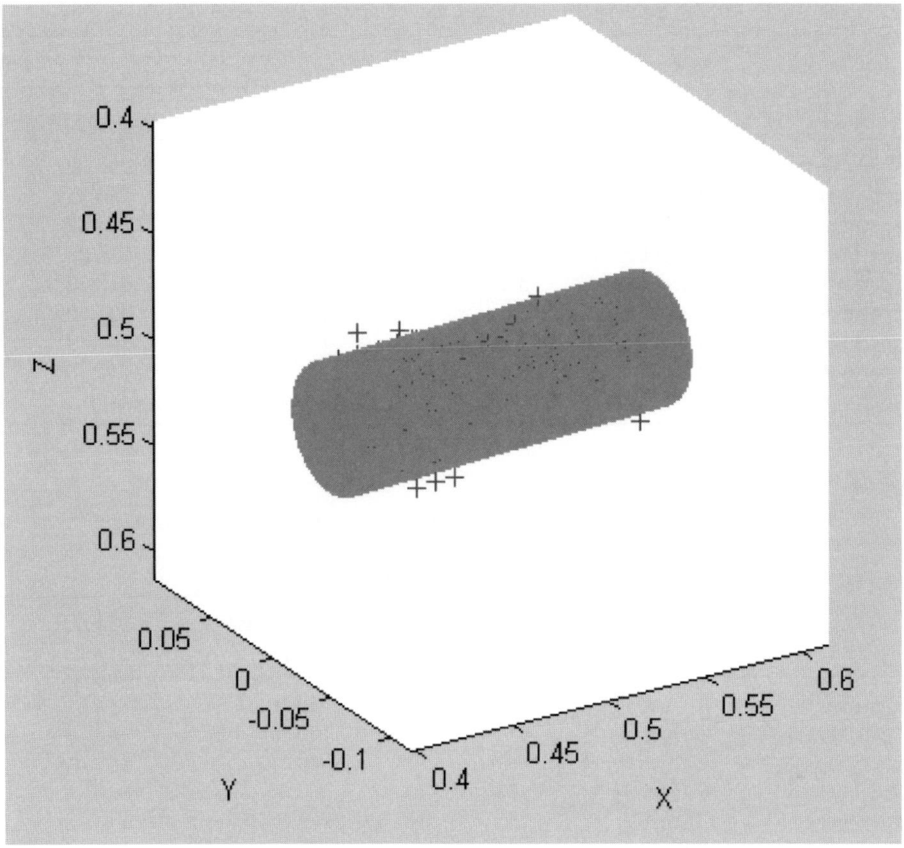

Figure 16.28. Cylinder fitted on the segmented data (Figure 16.26) according to the adopted model.

Table 16.2. Size parameters of the targets used in the tank experiments.

	Real values		
	Radius [cm]	Height [cm]	Volume [cm^3]
Target 1	3.35	10	352.5652
Target 2	2.9	10.5	277.4183
Target 3	2.55	8.1	165.4685
Target 4	3.5	8	307.8761

Table 16.3. Radius and height estimated values and relative volume computation.

	Estimated values			
	Target	Radius [cm]	Height [cm]	Volume [cm^3]
Experiment 1	Target 1	2.6	14	297.32
Experiment 3	Target 2	1.93	24	280.85
Experiment 4	Target 3	2.7	12	274.82
Experiment 5	Target 3	2.6	13	276.08
Experiment 6	Target 4	3.2	13	418.20
Experiment 7	Target 4	3.4	12	435.80
Experiment 8	Target 1	2.35	14	242.89
Experiment 9	Target 1	2.07	16	215.38
Experiment 10	Target 2	2.9	17	449.15

Table 16.4. Radius, height and volume estimation errors in percentage.

	% Errors			
	Target	Radius % error	Height % error	Volume % error
Experiment 1	Target 1	22	40	15
Experiment 3	Target 2	33	130	2
Experiment 4	Target 3	5	48	66
Experiment 5	Target 3	2	60	66
Experiment 6	Target 4	8	62	35
Experiment 7	Target 4	2	50	41
Experiment 8	Target 1	29	40	31
Experiment 9	Target 1	38	60	38
Experiment 10	Target 2	0	63	63
Average		*15.4*	*61.4*	*39.6*

16.3 CONCLUSIONS

The complete set of tools developed for 3D image processing proposed in Chapter 6 was tested on different sets of acoustic images. In particular, the availability of a simulator allowed us to carry out extensive tests on a large set of 3D images showing several objects (also complex in geometry) with different poses and viewpoints. Concerning the segmentation task, it was possible to evaluate algorithm performance by considering the degree of over- and under-segmentation of the results. The reconstructed models were compared with those expected, taking into account the spatial resolution of the sonar considered in the specific simulation.

To further assess the robustness of the developed methods, data from two different-scaled tank experiments were analysed. In both cases the results were satisfactory and showed the flexibility of the processing tool which, in different configurations of sonar systems and underwater scenes, was able to provide object reconstructions useful for a better understanding of the targets.

Part V

Evaluation

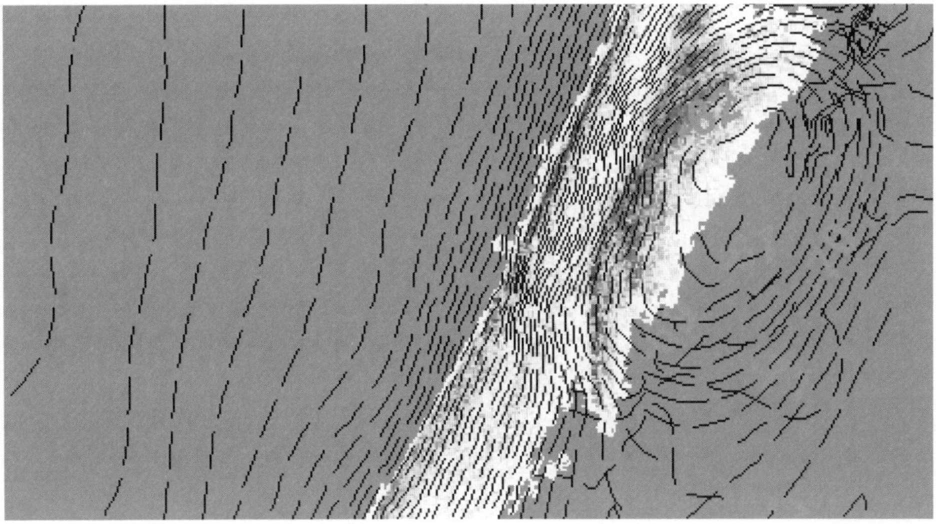

17

GIS-based data presentation system

A. Caiti, M. Robba, R. Sacile, P. Jonsson and L. Gelaziene

17.1 INTRODUCTION

The evaluation of a complex system like the SITAR data/integration presentation is not straightforward. In particular, from a software engineering point of view, one of the prerequisites of the final system is that of matching design specifications (which the system developed does respect). However, matching specifications cannot be the sole indicator for system evaluation. The final system must also get through subjective evaluation by the end-user community for which it was primarily developed. Subjective evaluation can be performed through a structured procedure and does indeed lead to important indications on the system as such and on future possible developments. In the following the subjective evaluation procedure followed for evaluation of the SITAR GIS-based data/integration presentation system and the results of the exercise are thoroughly described.

Before getting into the details of the evaluation exercise, it is remarked that some of the display capabilities and thematic map building of the system, integrating historical and SITAR-generated data, have already been shown in Chapter 7. The VRML display of reconstructed objects as presented in Chapter 16 (Figure 16.24) is fully integrated into the GIS system. Moreover, evaluation of the acoustic methodologies – in particular, the parametric side-scan sonar (described in Chapter 19) – will all be conducted using the geo-referenced maps built with the GIS system. Finally, in Figure 17.1 (see colour section) we show the thematic map of the "Stockholm gradient" described in Chapter 15 – in particular, the map of EROD-induced activity (see also Figure 15.12).

17.2 SYSTEM SUBJECTIVE EVALUATION

Subjective evaluation of the SITAR system was organized in the following way (see Figure 17.2 for the results): a panel of eight experts was assembled, with professional

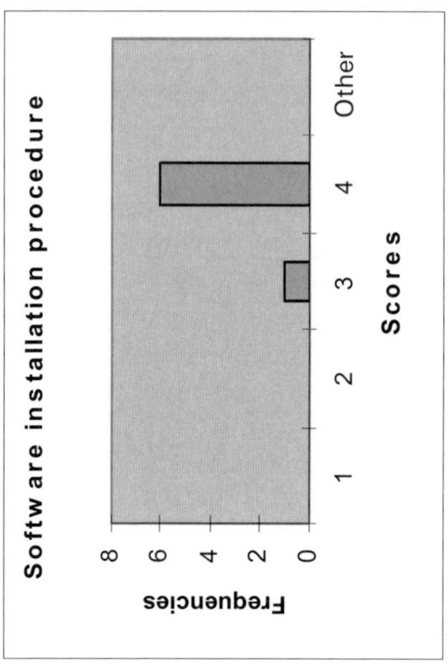
Software installation procedure
Frequencies
0
2
4
6
8
1
2
3
4
Other
Scores

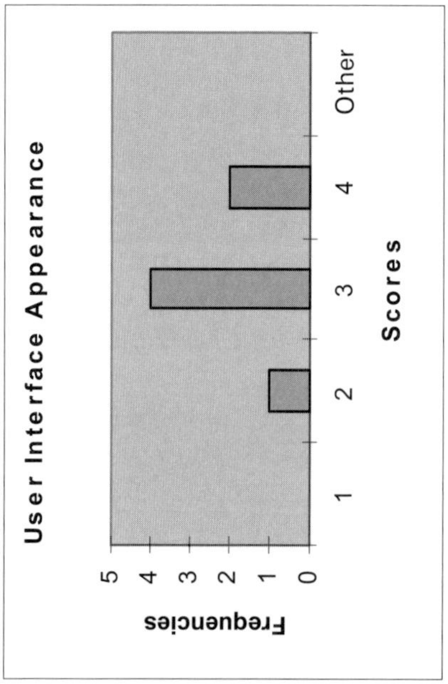
User Interface Appearance
Frequencies
0
1
2
3
4
5
1
2
3
4
Other
Scores

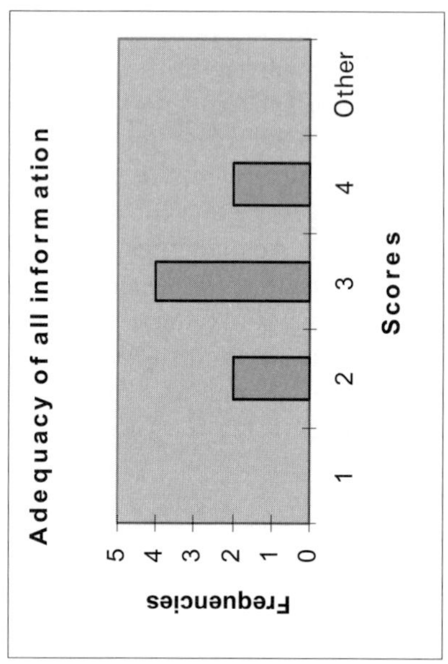
Adequacy of all information
Frequencies
0
1
2
3
4
5
1
2
3
4
Other
Scores

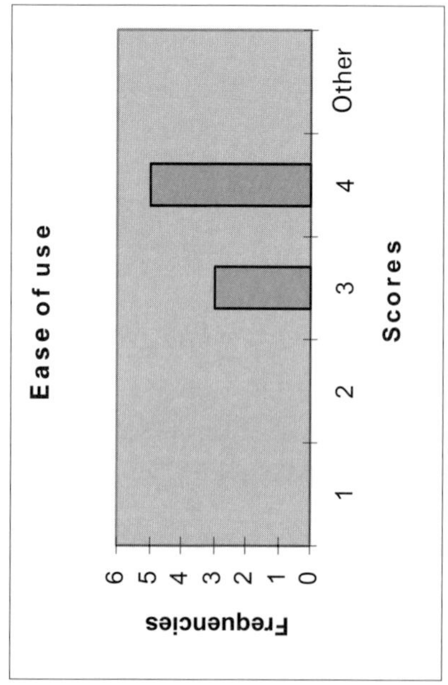
Ease of use
Frequencies
0
1
2
3
4
5
6
1
2
3
4
Other
Scores

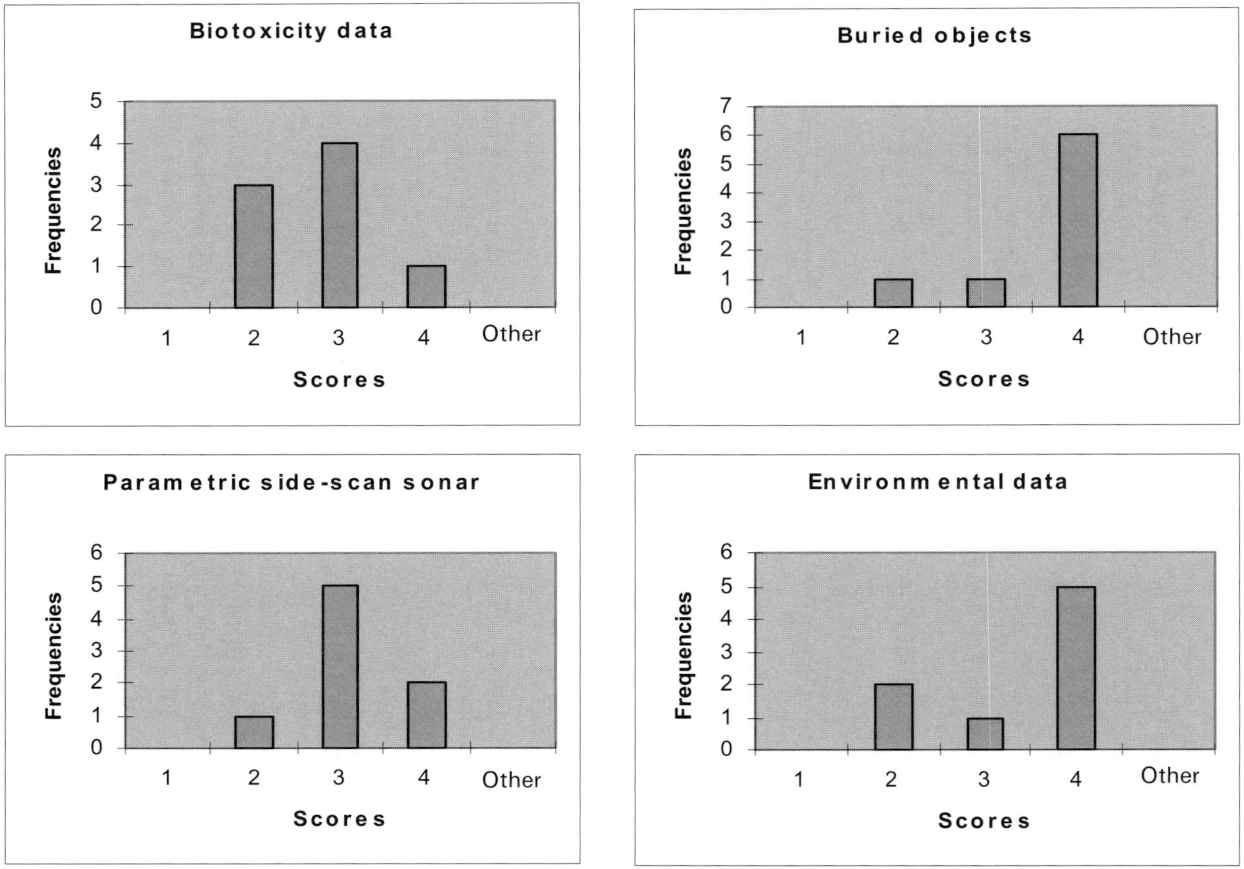

Figure 17.2. Results from the expert panel evaluation of the GIS-based data integration/presentation system. A score of 4 corresponds to "Excellent", and a score of 1 corresponds to "Poor".

competences ranging from environmental decision/policy making (three people), environmental science (two people), underwater acoustic and oceanic engineering (two people) and hydrographic survey (one person). The team was partly assembled with personnel from institutions within the project and partly with personnel from outside the project. After a short introduction to the system, each panelist was left with the system's CD-ROM, a user's guide and a PC, and were left to install the system and to play with it, exploring its different capabilities. At the end of the exercise, each panelist was asked to anonymously fill in a form to indicate her/his subjective judgement on the following eight topics:

(1) software installation procedure;
(2) user interface appearance (colours, organization of information, etc.);
(3) ease of use;
(4) adequacy of all the information presented (as related to risk assessment);
(5) biotoxicity data presentation;
(6) buried object presentation;
(7) parametric side-scan sonar presentation (standard and synthetic aperture);
(8) environmental data representation.

Judgement had to be given on each item according to the following score:

A = Excellent (score: 4)

B = Good (score: 3)

C = Sufficient (score: 2)

D = Poor (score: 1)

E = Insufficient or lacking (score: 0)

The scoring awarded to each item is reported as histograms in Figure 17.2 and as a "radar diagram" in Figure 17.3. Note that on some items some of the panelists did not award a score. The mode ("majority vote") on each topic is reported in Table 17.1. A clear indication of user satisfaction emerges from the subjective evaluation exercise.

From the histogram distributions, as well as the mode, the general satisfaction of the users with respect to the system can be inferred. However, the evaluation scores are best intended as a relative measurement, allowing some features of the system that are not as efficient as the others to be pinpointed. In particular, the results of the qualitative evaluation exercise seem to indicate that the user may still require some improvement on the presentation of toxicity and parametric side-scan sonar data, and on the global presentation for risk assessment need. The representation of standard environmental information and of buried objects from the multiple-aspect scattering technique was judged the best of all by the users. A focus group composed of the panelists and the ISME scientists in charge of software development was assembled to study the results of the evaluation and to summarize specific points that could be

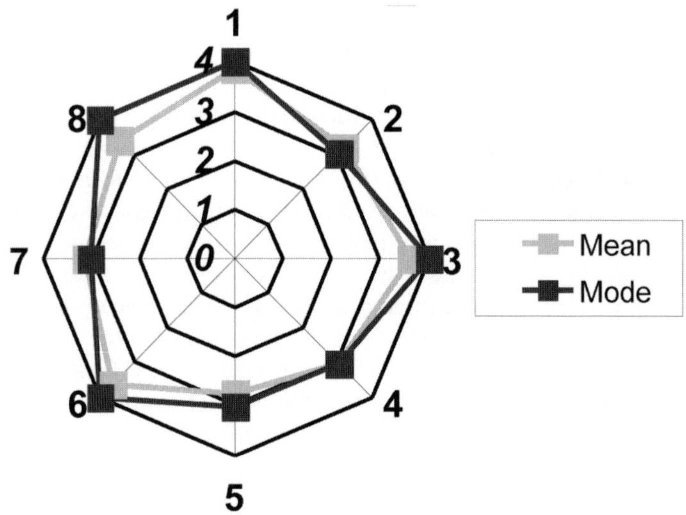

Figure 17.3. Overall results from the expert panel evaluation of the GIS-based data integration/presentation system. Scores are from the centre of the diagram toward the periphery. A score of 4 corresponds to "Excellent", and a score of 1 corresponds to "Poor". Evaluation items are from 1 to 8 clockwise. Light grey squares indicate the mean score per items, and dark grey squares indicate the mode (most voted score) per item.

further improved (and the means of doing so) in any future system design. A detailed list was reported in Caiti *et al.* (2004). While the overall qualitative evaluation has to be considered, in any case, very satisfactory, it has to be noted that some, if not all, of the suggested improvements in Caiti *et al.* (2004) could have been taken into account if included at some stage in the software specification. This was not possible because of the very nature of the data themselves: being part of project development, the data could not be anticipated *a priori*. This becomes obvious when one considers that the data classes awarded the best scores (mode: 4) are the standard environmental data

Table 17.1. Summary of the mode (most frequent score) on the eight selected topics for GIS system qualitative evaluation.

#	Topic	Mode of the score
1	Software installation procedure	4
2	User interface appearance	3
3	Ease of use	4
4	Adequacy of the information	3
5	Biotoxicity data presentation	3
6	Buried object presentation	4
7	Parametric side-scan sonar presentation	3
8	Environmental data representation	4

(on which it was easy for the users to specify their representation needs, since the users were familiar with the data) and the buried object data, for which the last stage of data processing and integration into the GIS system were the responsibility of two teams from the same partner (ISME), allowing immediate feedback from one system to the other and consequent software updates.

Evaluation of the effectiveness of the inclusion of SITAR-generated data for dumpsite risk assessment will be discussed in Part VI, in light of evaluation of the parametric side-scan instrument, multiple-aspect scattering methodology and biotoxicity measurements. Nevertheless, the qualitative evaluation results and the focus group discussion that followed indicated not only the satisfaction of the users in seeing SITAR-generated data together with standard data, but also the steering of future system design toward an even more relevant integration of SITAR-generated data. Hence, from this perspective on overall SITAR activity evaluation, it can be concluded that SITAR-generated data were evaluated as of great effectiveness and importance for environmental end-users in risk assessment of toxic dumpsites.

18

Biotoxicological methods

P. Jonsson

18.1 INTRODUCTION

The overall results from the biotoxicity test of the sediment samples in Möja Söderfjärd highlighted the absence of substantial polluting effects (see Chapter 15). These results could lead to at least the following interpretations of the results:

(1) the area is unpolluted and the method works;
(2) the area is polluted and the method does not work; and
(3) the area is polluted or unpolluted but the method gives irrelevant information.

Questions may arise whether (1) the sediment extraction procedure is relevant for the substances that may be released from dumping areas, (2) the analyses are reliable, (3) the injection procedure – including the extract doses – is biologically relevant. These questions will be discussed after the results have been put into a broader perspective below, by further making use of the available results from the so-called "Stockholm Archipelago gradient" (see again Chapter 15) and other similar studies.

18.2 THE STOCKHOLM ARCHIPELAGO GRADIENT

Since no differences in toxicity were found within the SITAR study area, six stations along a pollution gradient from Stockholm city centre to the open Baltic Sea were investigated using the same method (nanoinjection in newly fertilised rainbow trout, *Oncorhynchus mykiss*, eggs) for comparison (positive control). This gradient was chosen since toxic effects were documented there in adult feral perch (*Perca fluviatilis*). The data were reported in Chapter 15. The Stockholm Archipelago gradient showed gradually increasing sediment concentrations of ΣPCB andΣPAH from 2.5 μg/kg dry weight (dw) and 0.48 mg/kg dw in the outer archipelago to 143 μg/kg dw and 18 mg/kg dw, respectively, in Stockholm city centre. A similar distribution pattern was observed in perch muscle from the same stations (data not shown). EROD activity was low in the outer and central parts of the archipelago, including the dumping area, but increased significantly to high values in Stockholm city centre. Similar graded responses were obtained for total sediment extracts and for dicyclic and polycyclic aromatic fractions. The polycyclic aromatic fraction, however, had an EROD-inducing potency that was 3–11 times higher than that of the dicyclic aromatic fraction. EROD is one indicator among several other physiological (e.g., developing gonad frequency, somatic growth, gonadosomatic index, visceral fat somatic index) and biochemical (e.g., catalase, acetylcholine esterase, DNA adducts) variables that show that the environmental condition of perch is very poor in the Stockholm recipient.

The results obtained from the Stockholm Archipelago gradient show that the nanoinjection method is relevant for use in this type of semi-enclosed area. The low toxicity measured in Möja Söderfjärd and Möja Västerfjärd fits quite well into the overall pollution pattern of the Stockholm Archipelago with low concentrations of PCBs and PAHs in the outer and central archipelago. Also, the investigated biological/biochemical/physiological effects are insignificant in this part of the archipelago, indicating that no substantial large-scale releases of non-polar pollutants occur at the dumping area at present.

18.3 OTHER STUDIES USING THE NANOINJECTION TECHNIQUE

The nanoinjection technique has also been used in other investigations in Sweden. Some of the results are briefly reported below. In Örserumsviken Bay, which is heavily polluted by – *inter alia* – PCBs and PAHs from a paper mill, highly significant effects were demonstrated for mortality, different malformations and EROD activity when larvae were exposed to extracts of surficial sediment, settling particulate matter and semi-permeable membrane devices accumulating substances dissolved in the water (Sundberg *et al.*, 2003; Sundberg *et al.*, 2005a, b). Typical values for malformations induced by the most polluted sediments were 60–70% above control at a dose corresponding to 74-g dry sediment per kg egg. The corresponding EROD induction was 100 times the control value. These studies also included chemical fractionation, where differences in toxicity between fractions were seen.

In the Dalsland Kanal lake system significant effects were demonstrated for mortality, deformity and EROD activity in larvae exposed to extracts from some of these sediments (Balk *et al.*, 1995). At extract doses corresponding to 1–3-g dry sediment per kg egg, typical values for the highest effects were 20% above control for mortality and deformity and 200–400% of control for EROD activity.

The nanoinjection technique was also used to analyse the toxicity of settling particulate matter in the recipient (Kalmarsund) of the paper mill in Mönsterås (Åkerman *et al.*, 2005). At an extract dose corresponding to 12.5-g dry sample per kg egg, typical values for stations with the highest effects were 5–6% above control for mortality and deformity. AChE activity was inhibited 10–20%, whereas EROD activity was enhanced 3–4 times.

In one study concerning the potential toxicity of typical "urban city pollutants" (air particles, wood stove smoke, diesel exhaust, asphalt and tyres), their release to the aquatic environment of Stockholm city was investigated (Hansson *et al.*, 2004). Typical EROD values for larvae exposed to extracts of 50 mg per kg egg of the most potent samples (dry weight) ranged from 8–10 times that of control.

The toxicity of settling particulate matter in the aquatic environment of Stockholm was measured by Åkerman *et al.* (2005). Extract doses corresponding to 8–10-g dry sample per kg egg typically gave adverse effects of 15% above control for mortality and deformity and 1000–1500% of control for EROD activity. Fractionation demonstrated the differences in toxicity for the three fractions produced.

The results of the studies referred to above show that the nanoinjection technique can be used to demonstrate biological effects of a variety of non-polar pollutants in many different environments and matrices.

18.4 FINAL EVALUATION

Many methods are at hand for the measurement of the acute toxicity of pollutants in recipients. However, at least in Swedish waters, considerably fewer methods have been shown to describe long-term sub-lethal effects. The nanoinjection technique has proved to be useful in the detection of the effects of many types of non-polar pollutants in many types of matrices and may therefore be of great importance for inclusion in biological effect studies and environmental risk evaluation in both marine and freshwater systems. In the SITAR study area, no effects were found in the munitions dumping area, probably simply because no effects were to be found of release of non-polar pollutants from the munitions.

The method used within SITAR included chemical fractionation where different chemically defined fractions can be isolated. Only dicyclic and polycyclic aromatic fractions were studied, although many other fractions could have been investigated. The fractionation strategy makes it possible to detect which substance or group of substances have caused the effect. Subsequently, this may lead to measures to eliminate this substance from discharges. By using sediment cores that can be dated, samples may be taken from the same period of time in a gradient or in a grid covering the study grid. This makes it possible to perform retrospective studies of the historical development of the area from the pollution source.

It has to be mentioned that only non-polar extracts were studied in the biotoxicity test. Therefore, it cannot be excluded that biological effects may still be detected at the dumping area if a different extraction method is used; say, one that reveals the effects of water-soluble (polar) pollutants, which can also be analysed using the nanoinjection technique. Nanoinjection of aqueous solutions has been performed by Walker *et al.* (1996), Åkerman and Balk (1998), Åkerman *et al.* (1998) and Ekman *et al.* (2003, 2004).

Normally, sub-lethal effects develop slowly. The nanoinjection technique using rainbow trout eggs, however, may be slower than several other methods using other biological test matrices. One prerequisite of testing is that fresh rainbow trout eggs are available, which is the case only between January and May. (If salmon, *Salmo salar*, or sea trout, *Salmo trutta*, are accepted, eggs are available also in November and December.) This limitation in egg supply may cause up to half-a-year's delay if the decision to perform the study is taken shortly after the spawning season. In cases where a biological answer to a pollution discharge is needed relatively quickly, other methods with faster response times will have to be chosen.

One thing that hampers the use of this method by end-users of a more administrative persuasion is that it is difficult to interpret/calculate the exposure situation. One important question is how environmental concentrations correspond with the registered effects. To make it easier for the end-user, effects need to be related to the dose in a clearer way. In sediments most of the non-polar pollutants are related to the organic content or fractions of it. Several components of the organic matrix have been suggested for use in normalization of the concentrations of this type of pollutants – for example, total organic carbon (TOC), black carbon, organic lipids and soot particles. In different contexts one of them may be more suitable than the others. In any case it may be concluded that the dry weight of the sediment is not good for normalization in a study like this, since the sediment carbon content, which retains most of the non-polar compounds, may vary considerably between different recipients and along gradients. Therefore, it is suggested that the observed effects are related to TOC content. This will help the end-user to compare the effects observed in different areas.

19

Acoustical methods

M. Zakharia, Ph. Blondel, J. Dybedal, I. Karasalo, J. M. Hovem, A. Trucco, M. Robba and A. Caiti

19.1 INTRODUCTION

Evaluation of the results obtained by processing the acoustic data as described in Part IV will be now evaluated in light of the knowledge of the background conditions of the experimental tests. In this respect, SITAR tests were performed in two, rather different, situations: in laboratory tanks, with controlled conditions, and at sea, over an existing shallow water dumping area. While laboratory tests allows for a somewhat precise and systematic comparison between results and "ground truth", experiments at sea – in the SITAR case – allow comparison only with pre-existing knowledge on the site and with those portions of data for which experimental conditions were good enough to obtain final data of sufficient quality. As already mentioned, both the parametric side-scan sonar (PSSS) and the multiple-aspect scattering (MAS) technique operated in sub-optimal conditions in the Möja Söderfjärd experiment; on the other hand, the results obtained in the field test, though not as systematic as those obtained under laboratory conditions, corroborate the findings of laboratory tests, and give a field demonstration of the potential of the techniques investigated in SITAR. It has to be remarked that – in particular, for the PSSS – the evaluation, requiring comparison with historical data, was greatly facilitated by the availability of the GIS data presentation system: thanks to the GIS tool developed, we can easily establish a straightforward correspondence between the images obtained by the SITAR system and historical data (HF images, maps, sub-bottom profiles, video observations, etc.). In the following, the PSSS results are evaluated first, and then the MAS results are discussed.

19.2 LOCALIZATION WITH PARAMETRIC SIDE-SCAN SONAR

In Chapter 9 the results obtained in a full-scale tank experiment were reported and discussed, showing the merit of simultaneous display of co-located high- and low-frequency images, the capability of the system in localizing proud and submerged objects, and the gain obtained using synthetic aperture processing. In Chapter 13 the field data were discussed, showing again the advantage of synthetic aperture processing, and concentrating on the "bright spots" and "reverberating patterns" visible in the data. However, from the point of view of using the system for object localization in dumpsite risk assessment, the questions are: Do the "bright spots" correspond to objects on the seabed? Do the low-frequency images allow identification of buried features? The following discussion will show examples in which there is a positive answer to both questions.

In Figure 19.1 (see colour section) an enlargement of a PSSS run is shown (high-frequency data), together with the display of objects identified in previous surveys on the site by the Swedish Geological Survey. It is evident that "isolated bright spots" do correspond with the position of the object found by independent survey. The PSSS feature and the reported historical position of the object are about 5 m apart. However, in Figure 19.2 (see colour section) synthetic aperture processing of the same data is reported, showing that the single, isolated feature of Figure 19.1 may correspond to *two* different objects. Unfortunately, since this finding has become apparent only in processing, there is no immediate way of verifying this other than coming back to the site. It is not impossible, though, that in the historical data two close objects appeared as a single feature, and that subsequent visual inspection stopped when the first object was identified and inspected. Visibility at the bottom at the time of the SITAR trial was estimated at slightly more than a meter, and it is likely that the same conditions were also present when the Swedish Geological Survey conducted their inspection.

Isolated features showing only in the low-frequency data may be an indication of buried objects; however, since the low-frequency data were analysed only when back in the lab, it was not possible to collect ground truth at the positions of a potentially buried object; reverberating features, however, more likely associated with geological features may give an indication of the faithfulness of the PSSS data to the geological structure of the area and of the sub-bottom. In Figure 19.3 (see colour section) high-frequency data (110 kHz), low-frequency data (20 kHz) and low-frequency data after synthetic aperture processing are shown. A strong reverberating feature is clearly evident in the low-frequency data, with no correspondence in the high-frequency data (note also how synthetic aperture processing improves the continuity of the feature). If not an artefact, this feature must be associated with a buried geological structure. Geological ground truth is available in the form of a shallow seismic survey conducted in the area by the Swedish Geological Survey. Figure 19.4 (see colour section) shows the PSSS data (low frequency) and the position of the historical seismic lines. The closest available seismic transect is at a distance of about 50 m from the reverberating feature, close enough to give a reliable indication of the sub-bottom geological structure. The relevant portion of the seismic line is reported in Figure 19.5.

Figure 19.5 indeed shows the presence of a strong reflector at very shallow depths below the seafloor, which can be associated with the PSSS low-frequency reverberating feature.

A final comparison with available "ground truth" data is now made between the

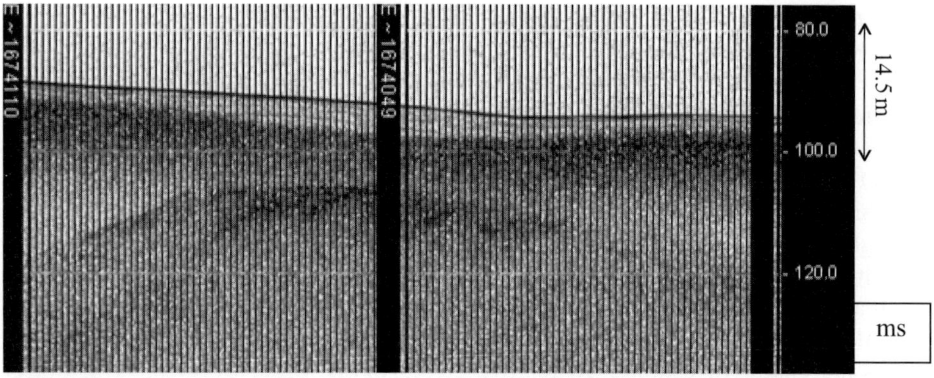

Figure 19.5. Sub-bottom profiling section ("ground truth" – historical data from Swedish Geological Survey) at closest range to the low-frequency reverberating feature in the PSSS. The right-hand side scale is in ms (two-way travel time). Depth conversion as indicated by the arrow was computed on the basis of constant sound speed in the media equal to 1450 m/s.

PSSS high-frequency data, in an area where a strong reverberating feature is present, and the data from the Swedish Geological Survey in terms of Klein standard side-scan sonar data and bathymetric data. The idea here is to verify whether the high-frequency data make available to the user the same kind of information that a user would like to have from standard side-scan sonar equipment. Comparison with the Klein data is reported in Figure 19.6 (see colour section). Although there are some similarities between the two acoustic images, visual interpretation suggests that there are also important differences that need explanation. However, when comparing the PSSS data with bathymetry (Figure 19.7, see colour section), it is quite clear that the PSSS data agree very well with the bathymetric data (and, to be frank, even the authors of this chapter were surprised to see such a good match). The same match does not happen between the Klein and the bathymetric data. The conclusion we draw is that the very good match between historical bathymetry and PSSS data shows that the PSSS does not perform less well in high-frequency resolution in the same way as standard side-scan sonars do. The poor performance of the Klein data with respect to bathymetry does not show in itself that the SITAR PSSS performs better than the Klein. One different explanation may be due in errors in geo-referencing the Klein data as made available to the SITAR team by the Swedish Geological Survey after mosaicking the different transects.

Evaluation of the PSSS in the field experiment may allow us to conclude that, despite sub-optimal configuration and experimental difficulties associated with the first sea test of the instrument, the system – together with the associated synthetic aperture processing – has the capability of identifying objects of the required dimension, of imaging sub-bottom features that can be clearly identified as buried by visual comparison between high- and low-frequency co-located data, of guaranteeing performance at least equal if not better than commercial state-of-the-art equipment when used in standard side-scan mode.

19.3 OBJECT INSPECTION USING MULTIPLE-ASPECT SCATTERING DATA

With respect to the PSSS system, evaluation of the MAS technique is more straightforward in the case of laboratory data, and more controversial in the case of field data. From the point of view of using the system for risk assessment evaluation of a seabed dumpsite, the question to be answered is: Does the system allow for inspection, through accurate reconstruction, of buried, man-made objects?

The richness of the dataset acquired by scaled tank experiments from the Bath Team allows for systematic evaluation of the errors in shape reconstruction obtained from the image analysis and feature extraction software package at the end of the processing chain. In Chapter 16 relative errors of the order of 30–65% in object dimension estimation were reported with objects of dimensions 5–10 cm. Considering the 10:1 scaling of the experiment, this amounts to a maximum 60-cm dimension reconstruction error in objects of dimension of 1 m, the minimum object size indicated as objective of the project. As for volume estimation, the errors reported are in the range 20–60%, with an average of about 40%, which is considered compatible with the assessment of toxic content in the barrels. The cases reported in detail in Chapter 16 referred to a buried and a half-buried cylinder. Quite surprisingly, there seems to be no relevant difference between buried and proud objects, with respect to the reconstruction error. The laboratory tests indicate that the set of procedures and algorithms investigated in SITAR is indeed able to reconstruct with adequate accuracy buried waste barrels.

Data collected in the field, due to equipment failures and experimental drawbacks, did not allow processing down to the end of the processing chain; moreover, only one object could be acoustically inspected within the experimental time window. This is in itself an indication that the experimental procedure devised for MAS, at least in the SITAR project, is too complex to be implemented as a standard survey routine for dumpsite inspection. On the other hand, SITAR's aim is not to produce a system or a set of operational devices, but to demonstrate the feasibility of the approach. From the point of view of their feasibility, both the acoustical 3D imaging approach and the inversion algorithm using the Ray–Kirchhoff numerical model produced results consistent with the ground truth as observed by video inspection. Again, field data do not allow for systematic conclusions; however, they do offer an indication that, even with a sub-optimal configuration, MAS data can provide enough information for estimation of some if not all the geometrical parameters of the illuminated object.

Part VI

Conclusions

20

Conclusions

A. Caiti

This book has presented a complete all-round picture of the scientific and technical activities of the SITAR team in the three years: January 2002–December 2004. Although the project included research from very diverse scientific disciplines, a common methodological framework was established among the scientists participating in the project, and it has been reflected in the structure of the book. From a set of methods, developed starting with the research state-of-the-art, through experimental testing in laboratory-controlled and also realistic at-sea conditions, results pertaining to the project objectives have been obtained and evaluated in light of the overall goal of the project: the development of enabling technologies and scientific methodologies for risk assessment of seabed dumpsites with buried waste containers. The conclusions that can be drawn, based in particular in the evaluation part of the book (Part V), are here summarized with respect to the specific objectives that motivated the work.

20.1 LOCALIZATION OF BURIED WASTE: PARAMETRIC SIDE-SCAN SONAR

The system has indeed shown great potential for localization of both buried and seafloor exposed features. Although experimental difficulties led to operation in sub-optimal configuration with respect to original design, the system has already demonstrated a performance superior to that of existing systems and good correspondence with ground truth data. Synthetic aperture processing has also shown, both in the field and in the tank experiments, to be feasible and very important – in particular, in the discrimination of clustered objects. A systematic investigation of the system performance – in terms, for instance, of detection rate and false alarm rate – is missing, and it will be indeed a non-trivial task to design an experiment to evaluate this performance in the field.

20.2 INSPECTION OF BURIED WASTE: MULTIPLE-ASPECT SCATTERING

Results from the tank tests were very successful, both in the collection of acoustic data and in the processing up to feature extraction, estimation of parameters as dimension and volume, and virtual rendering and visualization. The field test experienced several operational difficulties, and the entire processing chain could not be completed with field data alone. This leads to the conclusion that, while the bistatic MAS approach is indeed capable of capturing all the relevant information, experimental procedures have to be consistently reviewed before making this technique feasible for on-site survey.

20.3 ACCUMULATED BIOTOXICITY EVALUATION: THE NANOINJECTION TECHNIQUE

Despite the fact that – at the specific SITAR test site – no relevant toxicity levels were measured (which is in itself a positive fact, although for the project this may mean less spectacular and dramatic results), thorough testing and comparison with available datasets has shown that indeed the technique is a valid and sensitive way to assess *in situ* accumulated toxicity. One aspect to be considered when using this technique is the time required for the analysis: being based on the observation of larvae and fish development, progress over time is dictated by larval growth. This may not be a desirable fact when a fast assessment is needed; in this case, other techniques should be considered and/or investigated.

20.4 USER-ORIENTED DATA REPRESENTATION: THE SITAR GIS SYSTEM

The integration of environmental, acoustic and biotoxicological data in a data base system with a GIS interface, and with queries and interrogation defined according to end-users' specifications, has been very positively evaluated by the end-users themselves. The system has also been instrumental in allowing a fast and accurate comparison of SITAR results with available ground truth. While a system of this kind has to be considered a necessary requirement for any risk assessment evaluation, even in this case improved designs are still possible: quite typically in these kinds of developments, at the end of the development – when the potential of the system is fully realized – additional wishes and requirements are expressed by the end-user community. The experience of the SITAR interface may serve as a starting point for the design of a future system.

Overall, it can be fairly concluded that the scientific and technical approaches investigated in SITAR have successfully passed the "proof of concept" test that was the objective of the project. As in most research endeavours, this does not mean that the "problem is solved". There is still much work to do: on one side, toward transforming SITAR results into commercially available technological products, and, on the other side, toward improved and even better performing scientific methodologies. In order to make these further developments possible, the SITAR team actively involved a larger group of potential end-users over the three years of the project at various stages. This was done by organizing specific seminars and project presentations, by data sharing, by cooptation in some project exercises – as, for instance, in the case of GIS system evaluation. The awareness of the end-user community to the SITAR developments, coupled with the positive results of the project, will hopefully lead to systematic use of SITAR or SITAR-inspired results in the risk assessment of buried seafloor dumping.

As the writer of this concluding chapter was privileged with coordinating the efforts of the SITAR Team, he may be forgiven if, at the very conclusion of this book, he wishes to express his personal, whole-hearted gratitude and appreciation to all the colleagues involved in the project: in a multidisciplinary and complex investigation like SITAR, the fully cooperative attitude demonstrated by every member of the team, together with the scientific excellence in each respective field, have been key to the successful and timely conclusion of project activities.

References

CHAPTER 1

Federal/Länder Government Working Group "Chemical Munitions in the Baltic Sea" (1993) *Chemical munitions in the southern and western Baltic Sea. Compilation, Assessment and Recommendations*. Federal Maritime and Hydrographic Agency, Hamburg, Germany.

HELCOM CHEMU – the Ad Hoc Working Group on Dumped Chemical Munition (1994). *Report on chemical munitions dumped in the Baltic Sea*. Report to the 16th meeting of the Helsinki Commission. Danish Environmental Protection Agency, Copenhagen, Denmark, 38 pp.

Rapsch, H. J., Fischer, U. (2000) *Munition im Fischernetz – Atlasten in der Deutschen Bucht*, Isensee Verlag, Oldenburg [in German].

CHAPTER 2

Belkacem, A., Zakharia, M. E. and Besbes K. (2006) A fast algorithm for 3D planar synthetic aperture imaging. *Acta Acustica/Acustica*, **92**, 493–496.

Châtillon J., Bouhier, M. E. and Zakharia, M. E. (1992) Synthetic aperture sonar for seabed imaging: Relative merits of narrow-band and wide-band approaches. *IEEE J. of Ocean. Eng.*, January, **17**(1), 95–105.

Hovem, J. M. (2005) *Marine acoustics*, Applied Research Laboratory, Univ. of Texas, Austin (in press).

Special issue on synthetic aperture sonar, *IEEE J. of Ocean. Eng.*, January 1992, **17**(1).

Westervelt, P. J. (1963) Parametric Acoustic Array. *J. Acoust. Soc. Am.,* **35**, 535–537.

Zakharia, M. E. and Châtillon, J. (2002) Synthetic aperture mapping and imaging, in *Underwater Acoustic Digital Signal Processing and Communication Systems*, R. S. H. Istepanian and M. Stojanovic (eds), Kluwer Academic, Boston, pp. 37–88.

CHAPTER 3

Blondel, Ph., N. G. Pace, G. J. Heald and R. Brothers (2001) High-frequency bistatic scattering: Comparison of tank and sea experiments, in *Acoustical Oceanography*, T. G. Leighton, G. J. Heald, H. D. Griffiths and G. Griffiths (eds), *Proceedings of the Institute of Acoustics*, **23**(2), 276–282.

Blondel, Ph., McCloghrie, P., Pace, N. G., Heald, G. J. and Brothers, R. (2002) High-frequency bistatic bottom scattering: Modelling and experimental studies, in *Proceedings of the Sixth European Conference on Underwater Acoustics ECUA-2002, Gdansk, Poland*, pp. 21–29.

Canepa, G., Pace, N. G. and Pouliquen, E. (2002) Field measurements of bistatic scattering strength of a sandy seabed at 118 kHz. *Proceedings of the Sixth European Conference on Underwater Acoustics, Gdansk, Poland*, pp. 183–188.

Jayasundere, N. and Blondel Ph. (2004) *Multiple-Aspect Scaled Tank Experiments and Processing Scheme Evaluation*, SITAR Technical Report D-16, 49 pp. SITAR, Bath.

Karasalo, I. and Skogqvist, P. (2004) Transient scattering from submerged and buried objects. *Proc. European Conference on Underwater Acoustics 2004, Delft, Netherlands*, pp. 451–456.

Larsen, M. A. and Hovem, J. M. (2003) *Multiple-Aspect Scattering Measurement Requirements*, SITAR Technical Report D-10, 13 pp. SITAR, Bath.

Pace, N. G. (2004) Bistatic detection of buried objects, in *Underwater Acoustic Seminars from the SITAR Project*. SITAR, Livorno, Italy, CD-ROM, January.

Skogqvist, P. and Karasalo, I. (2003) A fast acoustic scattering method for buried objects, in *10th International Congress on Sound and Vibration, Stockholm, Sweden*, A. Nilsson and H. Bodén (eds), Vol. 5, pp. 2457–2465.

Williams, K. L. and Jackson, D. R. (1996) *A Model for Bistatic Scattering into Ocean Sediments for Frequencies from 10–100 kHz*. APL-UW Technical Report TR-9505, University of Washington.

Williams, K. L. and Jackson, D. R. (1998) Bistatic bottom scattering: Model, experiments and model/data comparison. *J. Acoust. Soc. Am.*, **103**(1), 169–181.

CHAPTER 4

Burton, A. J. and Miller, G. F. (1971) The application of integral equation methods to the numerical solution of some exterior boundary-value problems. *Proc. R. Soc. London, Ser. A*, **323**, 201–210.

Karasalo, I. (1994) Exact finite elements for wave-propagation in range-independent fluid-solid media. *J. Sound Vib.*, **172**, 671–688.

Saad, Y. and Schultz, M. H. (1986) A generalized minimal residual algorithm for solving nonsymmetric linear systems. *SIAM J. Sci. Stat. Comput.*, **7**, 856–869.

Skogqvist, P. and Karasalo, I. (2003) A fast acoustic scattering method for buried objects. *Proc. 10th Int. Conf. Sound and Vibration, July, Stockholm, Sweden.*

CHAPTER 5

Åkerman, G. and Balk, L. (1995) A reliable and improved methodology to expose fish in the early embryonic stage. *Mar. Environ. Res.*, **39**, 155–158.

Åkerman, G. and Balk, L. (1998) Descriptive studies of mortality and morphological disorders in early life stages of cod and salmon originating from the Baltic Sea. *Am. Fish Soc. Symp.*, **21**, 41–61.

Åkerman, G., Tjärnlund, U., Sundberg, H., Zebühr, Y., Broman, D. and Balk, L. (2002) *Environmental monitoring in the Municipality of Stockholm – Lake Mälaren and the Baltic Sea – Biology*. Report: sampling years 96/97, 97/98 and 98/99. Report to the Environmental Administration of Stockholm, 39 pp. [in Swedish].

Axelman, J. and Broman, D. (1999) Inventories and fluxes of polychlorinated biphenyls from a contaminated industrial site. *Environ. Toxicol. Chem.*, **18**, 1871–1881.

Balk, L., Tjärnlund, U., Ericson, G., Liewenborg, B. and Åkerman, G. (1995) *Investigations of sediment toxicity in the Dalsland channel lake system*. Miljö- och planenheten, Länsstyrelsen i Älvsborgs län. Report 1995:13 (ISSN: 1104-8271), 72 pp.

Balk, L., Liewenborg, B., Linderoth, M., Sundberg, H., Noaksson, E., Hansson, T., Tjärnlund, U., Hanson, M., Schiedek, D. and Åkerman, G. (2003a) Biomarker studies of female perch (*Perca fluviatilis*) in a chronically polluted gradient through the Stockholm archipelago. Abstract from the 2003 ICES Annual Science Conference, ICES CM 2003/M:08, pp. 118–119.

Balk, L., Liewenborg, B., Linderoth, M., Sundberg, H., Noaksson, E., Hansson, T., Tjärnlund, U., Hanson, M., Schiedek, D. and Åkerman, G. (2003b) Biomarker studies of female perch (*Perca fluviatilis*) in a chronically polluted gradient through the Stockholm archipelago. Paper from the 2003 ICES Annual Science Conference, ICES CM 2003/M:08, 7 pp.

Colmsjö, A. L., Zebühr, Y. and Östman, C. E. (1987) Group separation of PCDDs, PCDFs, PACs and aliphatic compounds on an amino bonded stationary phase for HPLC. *Chromatographia*, **24**, 541–544.

Ellman, G. L., Courtney, K. D., Andres Jr, V., Featherstone, R. M. (1961) A new and rapid colorimetric determination of acetylcholinesterase activity. *Biochem. Pharmacol.*, **7**, 88–95.

FLGWC (1993) *Chemical munitions in the southern and western Baltic Sea. Compilation, Assessment and Recommendations*. Federal/Länder Government Working Group, Federal Maritime and Hydrographic Agency, Hamburg, Germany, 60 pp.

Halsband, E. (1976) *The effect of seawater contaminated with N- and S-Lost on plaice, cormorant, and eel and of directly contaminated plaice and cod on guinea pig*. Bundesforschungsanstalt für Fischerei, Veröffentlichungen des Küsten- und Binnenfischerei, Hamburg, Nr. 60/1976, 33 pp. [in German].

Hansson, T., Åkerman, G., Tjärnlund, U., Grunder, K. and Balk, L. (2004) *PAH in Stockholm – Sources and Effects*. Investigation of sources with respect to biological effect, Prefinal Report 2004-11-01, Report to the Stockholm Water Administration, 24 pp. [in Swedish].

HELCOM CHEMU (1994) *Report on chemical munitions dumped in the Baltic Sea*. Report to the 16th meeting of the Helsinki Commission, *Ad Hoc* Working Group on Dumped Chemical Munition, Danish Environmental Protection Agency, Copenhagen, Denmark, 38 pp.

Henriksson, J., Johannisson, A., Bergqvist, P.-A. and Norrgren, L. (1996) The toxicity of organoarsenic-based warfare agents: *In vitro* and *in vivo* studies. *Arch. Environ. Contam. Toxicol.*, **30**, 213–219.

Lamparski, L. L. and Nestrick, T. J. (1980) Determination of tetra-, hexa-, hepta-, and octachlorodibenzo-p-dioxin isomers in particulate samples at parts per trillion levels. *Anal. Chem.*, **52**, 2045–2054.

MHLCAKL (1986) *Report on accumulation tests of war gas agents in fish.* MH 0052-28, Miljøstyrelsens Havforureningslaboratorium, Levnadsmiddelstyrelsen, and Civilforsvarets Analytisk-Kemiske Laboratorium, Danish Environmental Protection Agency, Copenhagen, Denmark, 36 pp. [in Danish].

Muribi, M. (1997) *Toxicity of mustard gas and two arsenic based chemical warfare agents on Daphnia magna.* Swedish Defence Research Establishment, Scientific report FOA-R–97-00430-222–SE (ISSN 1104-9154), 31 pp.

Muribi, M. and Eriksson, J. (1997) *Adsorption and toxicity of Clark I, Clark II and their metabolite tetraphenyldiarsine oxide.* Swedish Defence Research Establishment, Scientific report FOA-R–97-00527-222–SE (ISSN 1104-9154), 21 pp.

Näf, C., Broman, D. and Brunström, B. (1992) Distribution and metabolism of polycyclic aromatic hydrocarbons (PAHs) injected into eggs of chicken (*Gallus domesticus*) and common eider duck (*Somateria mollissima*). *Environ. Toxicol. Chem.*, **11**, 1653–1660.

Prough, R. A., Burke, M. D. and Mayer, R. T. (1978) Direct fluorometric methods for measuring mixed-function oxidase activity, in: *Methods in Enzymology*, Fleicher, S. P. L. (ed.), Vol. LII, Part C. Academic Press, New York, pp. 372–377.

Russian Federation (1993) *Complex analysis of the hazard related to the captured German chemical weapon dumped in the Baltic Sea.* National Report of the Russian Federation, CHEMU 2/2/1Rev. 1 (27 September 1993) as amended 11 October 1993, 31 pp.

Sundberg, H., Tjärnlund, U., Åkerman, G., Ishaq, R., Liewenborg, B., Zebühr, Y., Linderoth, M., Broman, D. and Balk, L. (2003) *Investigation of chemicals with biological activity in Örserum Bay*, Final Report March 2003. Report to the Municipality of Västervik, 28 pp. [in Swedish].

Sundberg, H., Tjärnlund, U., Åkerman, G., Blomberg, M., Ishaq, R., Grunder, K., Hammar, T., Broman, D. and Balk, L. (2005a) The distribution and relative toxic potential of organic chemicals in a PCB contaminated bay. *Mar. Pollut. Bull.*, **50**, 195–207.

Sundberg, H., Ishaq, R., Åkerman, G., Tjärnlund, U., Zebühr, Y., Linderoth, M., Broman, D. and Balk, L. (2005b) A bio-effect directed fractionation study for toxicological and chemical characterisation of organic compounds in bottom sediment. *Toxicol. Sci.*, **84**, 63–72.

Walker, M. K, Zabel, E. W., Åkerman, G., Balk, L., Wright, P. and Tillitt, D. E. (1996) Fish egg injection as an alternative exposure route for early life stage toxicity studies: Description of two unique methods, in: *Techniques in Aquatic Toxicology*, Ostrander, G. K. (ed.), CRC Press, Boca Raton, FL, pp. 41–72.

Zebühr, Y. (1992) *Trace analysis of polychlorinated dibenzo-p-dioxins, dibenzofurans and related compounds in environmental matrices.* PhD thesis, Department of Analytical Chemistry, Stockholm University, Sweden, p. 24ff.

Zebühr, Y., Näf, C., Bandh, C., Broman, D., Ishaq, R. and Pettersen, H. (1993) An automated HPLC separation method with 2 coupled columns for the analysis of PCDD/Fs, PCBs and PACs. *Chemosphere*, **27**, 1211–1219.

CHAPTER 6

Adams, R. and Bischof, L. (1994) Seeded Region Growing. *IEEE Transactions on Pattern Analysis and Machine Intelligence*, **16**(6), 641–647.

Alexandrou, D. and De Moustier, C. (1988) Adaptive Noise Canceling applied to Sea Beam Sidelobe Interference Rejection. *IEEE Journal on Oceanic Engineering*, **13**(2), 70–76.

Al-Fawzan, M. A. (2005) Methods for Estimating the Parameters of the Weibull Distribution, *InterStat, Statistics on Internet, http://interstat.stat.vt.edu/InterStat/intro.html-ssi*, last visited June 2005.

Baillard, C., Hellier, P. and Barillot, C. (2001) Segmentation of Brain 3D MR Images Using Level Sets and Dense Registration. *Medical Image Analysis*, **5**, 185–194.

Bowden, R., Mitchell, T. A. and Sarhadi, M. (1997) Real-time Dynamic Deformable Meshes for Volumetric Segmentation and Visualisation, in *Proc. British Machine Vision Conference, September, University of Essex, Colchester, Essex, UK*, Adrian F. Clark (ed.), Vol. 1, pp. 310–319.

Caiti, A., Murino, V., Palmese, M. and Trucco, A. (2003). Object reconstruction and feature extraction from 3-D underwater acoustic scattering measurements. *Proc. Tenth International Congress on Sound and Vibration, July, Stockholm, Sweden*, pp. 2449–2456.

Choi, S. M., Lee, J. E., Kim, J. and Kim, M. H. (1997) Volumetric Object Reconstruction Using the 3D-MRF Model-Based Segmentation. *IEEE Transactions on Medical Imaging*, **16**(6), pp. 887–892.

Giannitrapani, R., Trucco, A. and Murino, V. (1999) Segmentation of Underwater 3D Acoustical Images for Augmented and Virtual Reality Applications, *OCEANS '99, September, Seattle, WA*. MTS/IEEE, Seattle, pp. 459–465.

Malinverno, A., Edwards, M. H. and Ryan, W. B. F. (1990) Processing of SeaMARC Swath Sonar Data. *IEEE J. on Oceanic Engineering*, **15**(1), 14–23.

Murino, V. and Giannitrapani, R. (1999). Three-Dimensional Skeleton Extraction by Point Set Contraction. *IEEE Int. Conf. on Image Processing ICIP '99, October, Kobe, Japan*, Vol. III, pp. 565–569.

Murino, V. and Trucco, A. (2000) Three-Dimensional Image Generation and Processing in Underwater Acoustic Vision. *Proceedings of the IEEE*, **88**, 1903–1946.

Rodriguez, J., Loke, R. E. and du Buf, J. M. H. (2000) Fast Segmentation of 3D Data Using an Octree. *Proc. 11th Portuguese Conf. on Pattern Recogn., May, Porto, Portugal*, pp. 185–189.

Sauter, D. and Parson, L. (1994) Spatial Filtering for Speckle Reduction, Contrast Enhancement, and Texture Analysis of GLORIA Images. *IEEE J. on Oceanic Engineering*, **19**(4), 563–576.

Tek, H. and Kimia, B. B. (1997) Volumetric Segmentation of Medical Images by Three-Dimensional Bubbles. *Computer Vision and Image Understanding*, **65**(2), 246–258.

Umesh, P. S. and Chaudhuri, B. B. (2000) Region Based Techniques for Segmentation of Volumetric Histo-Pathological Images. *Computer Methods and Programs in Biomedicine*, **61**, 23–47.

CHAPTER 7

Breman, J. (2002) *Marine Geography GIS for the Oceans and Seas*. ESRI Press, Redlands, California, pp. 204.

Wright, D. and Bartlett, D. (1999) *Marine and Coastal Geographic Information Systems*. Taylor & Francis, London, pp. 320.

CHAPTER 8

Blondel, Ph., Pace, N. G., Heald, G. J. and Brothers, R. (2001) High-frequency bistatic scattering: Comparison of tank and sea experiments, in *Acoustical Oceanography*, T. G. Leighton, G. J. Heald, H. D. Griffiths and G. Griffiths (eds.), *Proceedings of the Institute of Acoustics*, **23**(2), 276–282.

Blondel, Ph., Dobbins, P. F. Jayasundere, N. and Cosci, M. (2006) High-frequency bistatic scattering experiments using proud and buried targets, in *Acoustic Sensing Techniques for the Shallow-Water Environment*, A. Caiti, R. Chapman, S. Jesus and J.-P. Hermand (eds). Springer, Dordrecht, pp. 155–170.

Dobbins, P. F. and Blondel, Ph. (2004) *Multiple-Aspect Scattering – Analyses of sea trials data*. SITAR Technical Report No. D-23, SITAR, Bath, 54 pp.

Dobbins, P. F., Blondel, Ph., Pace, N. G. and Karasalo, I. (2003) SITAR – Localisation and imaging of seafloor targets with multiple-aspect scattering, in *10th International Congress on Sound and Vibration, Stockholm*, A. Nilsson and H. Bodén (eds), Vol. 5, pp. 2467–2474.

Jayasundere, N. and Blondel, Ph. (2004) Multistatic Imaging of Multiple Targets – Scaled Tank Experiments with a Silt Seabed. *Proc. European Conference on Underwater Acoustics ECUA-2004, Delft, Netherlands*, 6 pp.

Karasalo, I. and Skogqvist, P. (2004) Transient scattering from submerged and buried objects. *Proc. European Conference on Underwater Acoustics ECUA-2004, Delft, Netherlands*, pp. 451–456.

Skogqvist, P. and Karasalo, I. (2003) A fast acoustic scattering method for buried objects, in *10th International Congress on Sound and Vibration, Stockholm*, A. Nilsson and H. Bodén (eds), Vol. 5, pp. 2457–2465.

CHAPTER 12

Beckmann, P. and Spizzichino, A. (1987) *Scattering of Electromagnetic Waves from Rough Surfaces*. Artech House, Norwood.

Berntsen, B., Hovem, J. M. and Bergem, O. (1999). Characterization of the seafloor using normal incidence acoustic backscattered time domain signals from a parametric sonar, in *Proc. Oceans '99 Conf., Seattle, WA*. MTS/IEEE, Seattle, pp. 30–36.

Berntsen, B., Frivik, S. A. and Hovem, J. M. (eds) (2000) *ISACS Final report*. Report to European Commission.

Eckart, C. (1953) The scattering of sound from the sea surface. *J. Acoust. Soc. Am.*, **25**, 566–570.

Hamilton, E. L. (1980) Geoacoustic modeling of the sea floor. *J. Acoust. Soc. Am.*, **68**, 1313–1340.

Widess, M. B. (1973) How thin is a thin bed? *Geophysics*, **38**(6), 1176–1180.

CHAPTER 14

Cantú-Pas, E. (1998) A Survey of Parallel Genetic Algorithms. Calculateurs parallèlles. *Reseaux es System Repartis,* **10**(2), 141–171.

Dobbins, P. F. and Blondel, Ph. (2004) *Multiple-Aspect Scattering: Analysis of Sea Trial Data*. SITAR Technical Report no. D-23, SITAR, Bath, 54 pp.

Nelder, J. A. and Mead, R. (1965) A Simplex Method for Function Minimization. *Computer Journal*, **7**, 308–313.

Storn, P. and Price, K. (1995) *Differential Evolution – A Simple and Efficient Adaptive Scheme for Global Optimization over Continuous Spaces*. ICSI, Berkeley, TR-95-012.

CHAPTER 15

Balk, L. (1985) *Characterization of xenobiotic metabolism in the feral teleost northern pike*. PhD thesis, Department of Biochemistry, Stockholm University, Stockholm, pp. 1–77.

Brewer, S. K., Little, E. E., DeLonay, A. J., Beauvais, S. L., Jones, S. B. and Ellersieck, M. R. (2001) Behavioural dysfunctions correlate to altered physiology in rainbow trout (*Oncorynchus mykiss*) exposed to cholinesterase-inhibiting chemicals. *Arch. Environ. Contam. Toxicol.*, **40**, 70–76.

Brunström, B., Broman, D., Dencker, L., Näf, C., Vejlens, E. and Zebühr, Y. (1992). Extracts from settling particulate matter collected in the Stockholm archipelago waters: embryolethality, immunotoxicity and EROD-inducing potency of fractions containing aliphatics/ monoaromatics, diaromatics or polyaromatics. *Environ. Toxicol. Chem.*, **11**, 1441–1449.

Fitzsimons, J. D., Huestis, S. and Williston, B. (1995) Occurrence of a swim-up syndrome in Lake Ontario lake trout in relation to contaminants and cultural practices. *J. Great Lakes Res.*, **21** (Supplement 1), 277–285.

Galgani, F. and Bocquené, G. (2000) Molecular biomarkers of exposure of marine organisms to organophosphorus pesticides and carbamates, in: *Use of biomarkers for environmental quality assessment*, Lagadic, L., Caquet, T., Amiard, J.-C. and Ramade, F. (eds). Science Publishers, Enfield, NH, pp. 113–137.

Gibson, G. G. and Skett, P. (1994) *Introduction to drug metabolism*, 2nd ed. Blackie Academic & Professional, London, pp. 88–93.

Hahn, M. E. (1998). The aryl hydrocarbon receptor: a comparative perspective. *Comp. Biochem. Physiol.*, **121C**, 23–53.

Okey, A. B., Riddick, D. S. and Harper, P. A. (1994) The Ah receptor: mediator of toxicity of 2,3,7,8-tetrachlorodibenzo-p-dioxin (TCDD) and related compounds. *Toxicol Lett.*, **70**, 1–22.

Russel, R. W, Gobas, F. A. P. C. and Haffner, G. D. (1999) Maternal transfer and *in ovo* exposure of organochlorines in oviparous organisms: a model and field verification. *Environ. Sci. Technol.*, **33**, 416–420.

Sundberg, H., Tjärnlund, U., Åkerman, G., Blomberg, M., Ishaq, R., Grunder, K., Hammar, T., Broman, D. and Balk, L. (2005a) The distribution and relative toxic potential of organic chemicals in a PCB contaminated bay. *Mar. Pollut. Bull.*, **50**, 195–207.

Sundberg, H., Ishaq, R., Åkerman, G., Tjärnlund, U., Zebühr, Y., Linderoth, M., Broman, D. and Balk, L. (2005b) A bio-effect directed fractionation study for toxicological and chemical characterization of organic compounds in bottom sediment. *Toxicol. Sci.*, **84**, 63–72.

CHAPTER 16

Palmese, M. and Trucco, A. (2004). A Flexible Method to Simulate 3-D Underwater Sub-Bottom Images. *Techno-Oceans 2004, Kobe, Japan*. MTS/IEEE, Kobe, pp. 2346–2353.

Murino, V. and Trucco, A. (2000). Three-Dimensional Image Generation and Processing in Underwater Acoustic Vision. *Proceedings of the IEEE*, **88**, 1903–1946.

V3D (2004) *3D Visualization Toolbox*, *http://www.mathworks.nl/matlabcentral/fileexchange*, last visited December 2004.

CHAPTER 17

Caiti, A., Gelaziene, L. and Jonsson, P. (2004) *Evaluation of data integration/presentation system*, SITAR Technical Report No. D-30. SITAR, Genoa, 16 pp.

CHAPTER 18

Åkerman, G. and Balk, L. (1998) Descriptive studies of mortality and morphological disorders in early life stages of cod and salmon originating from the Baltic Sea. *Am. Fish. Soc. Symp.*, **21**, 41–61.

Åkerman, G., Tjärnlund, U., Noaksson, E. and Balk, L. (1998). Studies with oxythiamine to mimic reproduction disorders among fish early life stages. *Mar. Environ. Res.*, **46**, 493–497.

Åkerman, G., Cato, I., Grunder, K., Broman, D. and Balk, L. (2005) *Environmental monitoring in the Kalmar Sound – Chemical analysis and biological effects*, Report: sampling years 2001/2002. Report to the Geological Survey of Sweden, 22 pp. [in Swedish].

Balk, L., Tjärnlund, U., Ericson, G., Liewenborg, B. and Åkerman, G. (1995). *Investigations of sediment toxicity in the Dalsland channel lake system.* Miljö- och planenheten, Länsstyrelsen i Älvsborgs län, Report 1995:13, 72 pp.

Ekman, E, Åkerman, G, Balk, L. and Norrgren, L. (2003). Nanoinjection as a tool to mimic vertical transmission of *Flavobacterium psychrophilum* in rainbow trout *Oncorhynchus mykiss*. *Dis. Aquat. Org.*, **55**, 93–99.

Ekman, E., Åkerman, G., Balk, L. and Norrgren, L. (2004) Impact of PCB on resistance to *Flavobacterium psychrophilum* after experimental infection of rainbow trout *Oncorhynchus mykiss* eggs by nanoinjection. *Dis. Aquat. Org.*, **60**, 31–39.

Hansson, T., Åkerman, G., Tjärnlund, U., Grunder, K. and Balk, L. (2004) *PAH in Stockholm – Sources and Effects, Investigation of sources with respect to biological effect,* Prefinal Report 2004-11-01. Report to the Stockholm Water Administration, 24 pp. [in Swedish].

Sundberg, H., Tjärnlund, U., Åkerman, G., Ishaq, R., Liewenborg, B., Zebühr, Y., Linderoth, M., Broman, D. and Balk, L. (2003) *Investigation of chemicals with biological activity in Örserum Bay*, Final Report March 2003. Report to the Municipality of Västervik, 28 pp. [in Swedish].

Sundberg, H., Tjärnlund, U., Åkerman, G., Blomberg, M., Ishaq, R., Grunder, K., Hammar, T., Broman, D. and Balk, L. (2005a) The distribution and relative toxic potential of organic chemicals in a PCB contaminated bay. *Mar. Pollut. Bull.*, **50**, 195–207.

Sundberg, H., Ishaq, R., Åkerman, G., Tjärnlund, U., Zebühr, Y., Linderoth, M., Broman, D. and Balk, L. (2005b). A bio-effect directed fractionation study for toxicological and chemical characterization of organic compounds in bottom sediment. *Toxicol. Sci.*, **84**, 63–72.

Walker, M. K., Zabel, E. W., Åkerman, G., Balk, L., Wright, P. and Tillitt, D. E. (1996) Fish egg injection as an alternative exposure route for early life stage toxicity studies: Description of two unique methods, in: *Techniques in Aquatic Toxicology*, Ostrander, G. K. (ed.). CRC Press, Boca Raton, FL, pp. 41–72.

Index

Printing: Mercedes-Druck, Berlin
Binding: Stein + Lehmann, Berlin